SPRINGER SERIES ON ENVIRONMENTAL MANAGEMENT

BRUCE N. ANDERSON

ROBERT W. HOWARTH

LAWRENCE R. WALKER

Series Editors

Springer Series on Environmental Management

The Professional Practice of
Environmental Management (1989)
R.S. Dorney and L. Dorney (eds.)

Chemicals in the Aquatic Environment:
Advanced Hazard Assessment (1989)
L. Landner (ed.)

Inorganic Contaminants of Surface
Water: Research and Monitoring
Priorities (1991) J.W. Moore

Chernobyl: A Policy Response Study
(1991) B. Segerståhl (ed.)

Long-Term Consequences of Disasters:
The Reconstruction of Friuli, Italy, in its
International Context, 1976-1988 (1991)
R. Geipel

Food Web Management: A Case Study
of Lake Mendota (1992) J.F. Kitchell
(ed.)

Restoration and Recovery of an
Industrial Region: Progress in Restoring
the Smelter-Damaged Landscape near
Sudbury, Canada (1995) J.M. Gunn (ed.)

Limnological and Engineering Analysis
of a Polluted Urban Lake: Prelude to
Environmental Management of
Onondaga Lake, New York (1996)
S.W. Effler (ed.)

Assessment and Management of Plant
Invasions (1997) J.O. Luken and
J.W. Thieret (eds.)

Marine Debris: Sources, Impacts, and
Solutions (1997) J.M. Coe and
D.B. Rogers (eds.)

Environmental Problem Solving:
Psychosocial Barriers to Adaptive
Change (1999) A. Miller

Rural Planning from an Environmental
Systems Perspective (1999) F.B. Golley
and J. Bellot (eds.)

Wildlife Study Design (2001)
M.L. Morrison, W.M. Block,
M.D. Strickland, and W.L. Kendall

Selenium Assessment in Aquatic
Ecosystems: A Guide for Hazard
Evaluation and Water Quality Criteria
(2002) A.D. Lemly

Quantifying Environmental Impact
Assessments Using Fuzzy Logic
(2005) R.B. Shepard

Richard B. Shepard

Quantifying Environmental Impact Assessments Using Fuzzy Logic

With 42 Illustrations

 Springer

Richard B. Shepard
Applied Ecosystem Services, Inc.
Troutdale, OR 97060
USA
rshepard@appl-ecosys.com

Series Editors:
Dr. Bruce N. Anderson
Planreal Australasia
Keilor, Victoria 3036
Australia
bnanderson@compuserve.com

Dr. Robert W. Howarth
Program in Biogeochemistry
 and Environmental Change
Cornell University
Corson Hall
Ithaca, NY 14853
rwh2@cornell.edu

Dr. Lawrence R. Walker
Department of Biological
 Sciences
University of Nevada
 Las Vegas
Las Vegas, NV 89154
walker@unlv.nevada.edu

Cover illustration: Fig. 9.18, page 94. Intersection, conjunction, T-norm, minimum.

Library of Congress Cataloging-in-Publication Data
Quantifying environmental impact assessments using fuzzy logic / Richard B. Shepard.
    p. cm. — (Springer series on environmental management)
    Includes bibliographical references (p.) and index.
    ISBN 0-387-24398-4
    1. Environmental impact analysis. 2. Fuzzy logic. I. Shepard, Richard B. II. Series.
    TD194.6.Q36 2005
    333.71′4—dc22   2005040214

ISBN-10: 0-387-24398-4   Printed on acid-free paper.

Printed in the United States of America.   (SBA)

9 8 7 6 5 4 3 2 1   SPIN 10979020

springeronline.com

Preface

Formal requirements for the assessment of environmental impacts of development activities may have begun in the United States with the passage of the National Environmental Policy Act (NEPA) in 1969, but they now are found in more than 200 countries worldwide. The details vary, but the underlying goals of minimizing environmental degradation and improving environmental conditions are the same. In many countries, these national requirements are supplemented by additional requirements by states, provinces, counties, cities and other political divisions that are collectively called sub-national statutes and regulations.

I am most familiar with the requirements at the national level in the United States as well as at the state and county levels in the western half of the country. However, books, other published reports, and communications with peers amply document that problems caused by the subjective nature of environmental impact assessments are international in scope. This subjectivity can be quantified and treated with mathematical rigor by the application of advanced computational intelligence techniques. This approach will work equally well regardless of geographic location or political jurisdiction because it is responsive to variations in societal values, legal frameworks, and regulatory agency practices. Trans-national organizations such as the European Union, World Bank, and United Nations Environmental Program also set project environmental standards that must be met in addition to the standards set by local governments.

It is important to make this disclaimer emphatically: in no way is this book to be taken as criticism of environmental impact assessment (EIA) laws, regulations, practitioners, or theorists. Such criticism would be neither warranted nor justified. Identifying subjective aspects is not criticism. Such identification forms the basis for understanding this book and benefitting from its content.

This book has three objectives:

1. The first objective is to document how environmental impact assessments have been conducted and to explain when and why contention develops. The book stresses that environmental assessments (whether of impacts or of existing conditions) are subjective expressions of societal, group, and individual values and opinions. As such, they are not objective or measurable. Science, particularly ecology and environmental science, has difficulties dealing with feelings, beliefs, and values, which are "nonscientific" concepts.

The specifics of the EIA process vary with the controlling jurisdiction; there is no attempt to describe all the variations and subtle differences, because this is not a book to teach the theory and detailed practice of environmental impact assessments as implemented worldwide. However, specific points will be based on my experiences as well as what others have experienced and described in published literature.

The first objective establishes two important points: EIAs are subjective and existing assessment methods do not effectively handle subjectivity. We speak and write using terms that cannot be measured. Concepts such as *significant*, *distant*, *acceptable* and others are understood by everyone – but we each may have a different definition of these terms. Almost all environmental regulatory processes depend on such imprecise, vague, inherently uncertain terms. Commercial development may be prohibited on *steep* slopes, but how is *steep* defined? It is almost always an arbitrary, crisp value; for example, 20 percent. This does not mean that a slope of 19.5 percent is not *steep*, but it means that there is no sharp threshold that separates *steep* from *not steep*. Fortunately, fuzzy sets, fuzzy logic, and approximate reasoning (among other computational intelligence methods) handle subjectivity effectively by quantifying it and manipulating it with mathematical rigor.

2. The second objective is to justify the use of fuzzy sets, fuzzy logic and approximate reasoning to provide decision-makers with the ability to make well-informed decisions: ones that are technically sound and legally defensible. I do this by describing core issues of an environmental impact assessment in terms of fuzzy modeling and other computational intelligence techniques.

   The concept of fuzzy sets was developed explicitly to address the inherent imprecision of everyday language which we all use to express ideas that cannot be measured. Fuzzy logic is the mathematics that permits rigorous operations on fuzzy sets to arrive at a outcome that is meaningful and can be explained. Approximate reasoning is the computer modeling of how humans make decisions (IF this THEN do that) when all the input data are subjective and not directly measurable. Other advanced techniques of artificial intelligence (AI) (including expert systems, decision support systems, and data mining using neural networks and evolutionary algorithms) also can be effectively and productively applied to addressing the underlying purposes of environmental impact assessments.

3. The third objective is to illustrate the use of computational intelligence techniques presented in objective 2 for environmental impact assessment. This example creates an approximate reasoning model applied to a project completed the traditional way under Washington state laws and regulations. While the example is based on a real industrial development proposal, the original environmental impact assessment was not developed with computational intelligence techniques. Therefore, the example has been adapted to demonstrate the application of these tools by adding missing information and deleting some components to make the example a reasonable size.

Mathematical symbols used in fuzzy logic (from [9]).

Symbol	Meaning
$\neg$	set NOT (also complement or inversion)
$\cap$	set AND (also intersection operator)
$\cup$	set OR (also union operator)
$\aleph$	higher-dimensional fuzzy space
[x,x,x]	fuzzy membership value
$\in$	member of a set; within
poss(x)	the possibility of event x
prob(x)	the probability of event x
{x}	crisp, or Boolean, membership function
$\bullet$	dyadic operator
$\xi(x)$	expected value of a fuzzy region
μ	fuzzy membership function
$\propto$	proportionality
$\mu(x)$	membership, or truth, function in fuzzy set
$\Re$	element from domain of fuzzy set
$\otimes$	Cartesian product or space
$\oslash$	empty, or null, set
$\supset$	implication
$\wedge$	logical AND
$\vee$	logical OR
Σ	summation

Acknowledgments

Among all the people whose efforts have brought me to the level of understanding and experience that allows me to write knowledgeably about environmental impact assessments and fuzzy system models a handful stand out of the crowd. Earl Cox introduced me to fuzzy sets and fuzzy logic with the first edition of his *The Fuzzy System Handbook* a decade ago. Since then his comments and suggestions have been helpful to my understanding of this subject and the subtleties of writing software that function to compute fuzzy system models with parallel rule-firing and the ability to solve otherwise intractable problems. Dr. William Siler created a parallel-firing fuzzy inference engine that is at the core of the solutions presented here. Three other friends and professional colleagues deserve public thanks for the highly productive conversations we have had over the years on environmental issues and running natural resource industries. These friends, Jonathan Brown, Paul Scheidig, and Ivan Urnovitz, manfully read drafts of the book and let me know at what sections their eyes started to glaze over. Paul Scheidig bravely read the first two parts and gave me his usual invaluable feedback. The suggestions from all the above for elucidation and clarity make this book a much better work. My fiancee, Pamela Sue Alexander, cheerfully accepted my long hours at the computer with patience and understanding; this made the process both easier and more pleasant. My editor at Springer, Janet Slobodien, has been a great guide into the world of book publishing, a world much different from that of peer-reviewed scientific or trade journal publishing. Despite the best efforts of all these outstanding people, any errors or mistakes that remain are mine alone.

Richard B. Shepard
February 2005
Troutdale, Oregon

Contents

List of Figures

List of Tables

1

Introduction

Human activity has been altering—for better and for worse—natural environments for thousands of years. Beginning several decades ago societies decided that they wanted a better understanding of human interactions with other species and the abiotic environment. Out of this concern have come laws and regulations requiring the evaluation of potential environmental impacts caused by continuing and new human activities. While science plays a role in the evaluation, the criteria by which activities are judged are those that reflect the societal values of the regulating jurisdiction. These governments, and other environmental impact assessment (EIA) professionals, need practical and effective tools in order to be effective.

There is a hierarchy which can be used to organize the practical application of these tools. At the top of this hierarchy is environmental risk management. Beneath that are levels of environmental management systems, regulatory oversight and approval, compliance assurance, violation enforcement systems and reclamation activities (which include remediation, enhancement and creation, among others). Environmental risk management is a process to guide businesses through the potential uncertainties of environmental laws and regulations. This includes bringing all their facilities and operations into compliance with existing regulations, maintaining full compliance, providing as much certainty and predictability as is possible under current conditions and planning for change. Governments, too, manage environmental risk through their statutory and regulatory programs. Environmental management systems are the detailed implementations of the environmental risk management policies. Environmental impact assessments are usually part of the lower hierarchical levels, implemented by national, regional and local governments.

Environmental impact assessment have been statutory and regulatory requirements for businesses around the world for more than three decades. In the United States these assessments are required under the National Environmental Policy Act (NEPA) in the form of Environmental Assessments (EA) and Environmental Impact Statements (EIS). At the state level assessments may also be required regardless of whether a major federal action that triggers NEPA compliance is involved. Washington State has its State Environmental Act (SEPA), Oregon has its Statewide Planning Goals (in particular, Goal 5: Natural Resources, Scenic and Historic Areas. and Open Spaces) and California has its California Environmental Quality Act (CEQA). Other states have their environmental assessment laws and regulations, too. These regulations apply to industry (e.g., mining, forestry, electrical generation/transmission), commerce (e.g., transportation) and development by ports, resort operators, housing builders and others. In other countries there are equivalent national and subnational requirements for development and industrial projects.

No matter where the project is located or what it involves, there are a number of EIA characteristics and procedures with the potential to frustrate the various interests involved in the process. Too often the frustration produces contention which increases EIA cost and time without resolving or reducing the underlying concerns of all parties. One of the most important characteristics which creates the potential frustration elements is the inherently subjective nature of environmental impact assessments. Because societal values of significance and acceptability are impossible to measure objectively, decisions may seem arbitrary and capricious to groups who do not like the outcome. Much too frequently, these decisions end up in the legal system, where costs and time delays continue to accumulate while the subjectivity is not removed.

There are also real problems that arise during the environmental impact assessment process, and these may also lead to delays, increased costs, and dissatisfaction by one or more s in the process. Examples of real problems are—

- Quantitatively describing existing conditions (environmental, social and economic) so they can be compared with alternative outcomes.
- Establishing the relative importance of various condition components so the most important ones are included in the assessment.
- Ensuring the equal and objective analysis of each alternative.
- Having a technically sound and legally defensible basis for making a decision based on the assessments.

Several of these potential problems—very important ones—will be thoroughly documented in Part I of this book. Real problems and "frustration elements" may be related in any specific assessment process.

A method of solving the identified problems, a mathematically rigorous, alternative environmental impact assessment process using fuzzy logic and approximate reasoning, will be described and justified in Part II. To illustrate how such a computational intelligence approach would work in performing an assessment, an example is presented in Part III. The computer modeling system used for the example in Part III is FuzzyEI-Assessor™. Other fuzzy system modeling software is available at no cost or can be licensed from the developers. These could be used to implement the modern approach described in Part II.

Two factors differentiate the approach using fuzzy system models from those techniques used in the past:

1. It takes advantage of advanced computational intelligence techniques (such as fuzzy sets and logic) for quantifying and manipulating (in a mathematically rigorous way) subjective, inherently uncertain or imprecise values and concepts.
2. While the protocol is widely applicable and valid under many different environmental statutes, regulations, and societal values, its application is site specific but consistent.

In addition, the process works much more effectively with the active input of s, the public, and everyone else with an opinion. The approach to conducting EIAs described in this book is technically classified as *multi-objective, multi-criteria decision-making under conditions of uncertainty*. It is part expert system, part decision support system, and all well supported in set theory, operations research, decision science, ecosystem ecology, and other disciplines.

1.1 Making Decisions

There is a subset of operations research called "decision science" that focuses both on how decisions are made and on how to make better decisions. Decision science also incorporates elements from psychology, business, computer processing of information, and other interdisciplinary aspects of importance to managers. Many of the ideas and techniques are appropriate to the environmental impact assessment process. After all, the underlying purpose of conducting an environmental assessment is to create a basis for making informed decisions on the condition or acceptability of a natural ecosystem (e.g., the status of Coho salmon populations along the Pacific Ocean coast of Oregon), whether to allow an industrial or commercial project to be developed, or to determine the appropriate alternative for such development. This book focuses on identifying the alternatives among which a decision will be made and predicting the significance of the future environments under each alternative.

1.1.1 Decision Support Systems

There are two components involved in making a decision [23]:

1. Judgment. The experience, feelings, and insight that the decision-maker uses consciously or unconsciously.
2. Knowledge. The project-specific information that has been collected and analyzed for use in making an informed decision.

The environmental impact assessment process is supposed to be based on knowledge of the specific environment and similar projects undertaken in the past. Judgment—including the "best professional judgment" of the technical staff—is always a factor. The goal of the process should be a result that is informed and defensible. Several factors make it difficult to reach this goal, including having multiple objectives, the presence of many (often conflicting) constraints to be satisfied, and the accumulation of a large amount of project-specific information to be properly analyzed and understood in the context of the project. This last factor reflects the many scientific, economic, social, and political considerations that cannot be ignored in the decision-making process. Within the broad category of decision science, environmental impact assessments are in the broad category of *multi-objective, multi-criteria* decisions.

1.1.1.1 Multi-Objective

Multi-objective means that there are different goals to be attained. For example, operation of a dam and reservoir system may need to be optimized for flood control, irrigation, barge traffic, hydroelectric power production, and recreational use. Land development at a port may need to be optimized for heavy industry, light industry, maritime, rail and truck shipping, cargo storage, wildlifeprotection, and general economic gain. These objectives can sometimes seem to be mutually exclusive, but with the proper techniques and approaches an optimal balance can almost always be achieved. This balance maximizes all objectives relative to each other. Linear programming (out of the field of operations research) is a long-standing example of multi-objective optimization in decision-making and is sometimes taught in graduate courses of systems ecology.

1.1.1.2 Multi-Criteria

Multi-criteria describes the different values that constrain achievement of the identified objectives. In operations research multi-criteria decisions result from the use of a "bottom up" approach when the values of different public groups shape the outcome. Criteria for a project might include increasing the number of available jobs, providing a high rate of return on investment, protecting the environment, increasing the size of a population of fish or wildlife, and being æsthetically pleasing.

In addition to having multiple objectives and being constrained by multiple criteria, the EIA process occurs under conditions of future uncertainty and frequently involves more than one decision-maker. This makes it very difficult to reach the appropriate decision; one that can be satisfactorily explained and completely justified. Historically, deciding on a decision method emphasized the psycho-social interactions of the group and how it reached consensus. The powerful computational intelligence tools now available bring advantages to all levels of decision-making. Among these advantages are their ability to quantify subjective values, beliefs, and language and the ability to minimize, or eliminate, qualitative pressures from the decision.

Several mathematical approaches to handling imprecise concepts have been developed and used in different situations. Each approach has a type of application where it works very well. Environmental impact assessments are characterized by the importance of *linguistic variables* that are not directly measurable (that is, they are terms easily expressed in language that are not represented by discrete numbers or thresholds). Fuzzy sets are used to convert the subjective linguistic variable into a number and fuzzy logic manipulates these numbers in a mathematically rigorous way.

The above overview describes what is required by a decision support system (DSS). Environmental impact assessment are a form of DSS, so the approach used to conduct the EIA should accommodate multiple objectives and multiple criteria under conditions of uncertainty. Over the past few decades the power of computers has been applied to the complexity of decision-making. Computers are far more capable than people at making sense out of large amounts of data. There are many different decision support system applications available to the business community, and it is certainly reasonable to demand that such an application facilitate decisions when assessing environmental impacts of a proposed project.

However, people are far superior to computers when it comes to judgment and the extrapolation of past experience to new situations. Human expertise is another facet of environmental impact assessments that contributes to informed and legally defensible decisions.

1.1.2 Expert Systems

The capture of an expert's knowledge within computer software was one of the first artificial intelligence (AI) efforts in the 1960s and 1970s. These knowledge-based expert systems, or expert systems for short, capture what an expert would do in a particular situation as a series of IF-THEN rules. While in many ways the promise of expert systems was over-hyped, they are commonly used today in fields as diverse as medical diagnosis, psychiatric screening, new product pricing, insurance fraud detection, financial portfolio evaluation and other business processes [8].

During the EIA process technical experts are employed to analyze, interpret and explain the large amount of technical data that describe natural ecosystems. More importantly, IF-THEN "rules" are precisely how alternatives are analyzed, but not in a structured, formal way. For example, the EIA report might include statements such as, "if the wetland is filled, then migratory waterfowl habitat will be decreased" or "if the 'No Action' alternative is selected, then 80 high-paying, permanent jobs will not be created." Translating the project-specific IF-THEN rules into a form that can be used by a computer permits these rules to be uniformly applied to all alternatives. When human reasoning is emulated by computer software, the result is called an approximate reasoning model. Because this software applies the collective knowledge of experts to solve the problems of environmental impact assessments the application is also an expert system.

What the appropriate expert system model provides to the decision-maker is an objective characterization of the baseline environmental conditions along with projected environmental characterizations of the future conditions when each project alternative is applied. These alternative characterizations are calculated with input of all stakeholders on values such as "acceptability," "significance," "sustainability," "biodiversity," and other such concepts that cannot be objectively measured. This makes the process a true decision support system rather than a decision-making system.

1.1.3 Decision-Making in the EIA Process

Environmental decision-making involves four broad steps:

1. Problem identification.
2. Options (alternatives) identification.
3. Predictions of the future environments associated with each alternative.
4. Selecting one of the options.

Problem identification is almost always defined under a law at the national, regional, or local level and will not be further discussed. The next two steps (alternative identification and predictions of potential futures) *are* the focus and purpose of this book. The final component—selecting one alternative action—almost always is based on social values, economic priorities, and political considerations and will not be further considered. However, if the identification and characterization of alternative options is done with sufficient scientific objectivity and mathematical rigor then the selection of a particular option should be simplified—even in the most sensitive and contentious situations.[1]

The decision-making process can be further organized into the categories of information gathering and analyses. These categories include—

- Identifying values.
- Characterizing (not only describing) the existing environment.
- Characterizing the social, economic and political setting.
- Characterizing the legal and regulatory setting.
- Integrating information.
- Forecasting effects of project alternatives.
- Assessing options.
- Post-decision assessment and justification.

Various tools, methods, and techniques have been applied to these categories. This book will address the needs of decision-makers in most of environmental impact assessment components.

---

[1] This is not to assume naively that everyone will be pleased with the outcome, but the steps leading to the decision will be visible and shown not to be arbitrary and capricious.

Part I

The Traditional Approach

2

General Principles

> Impact assessment, simply defined, is the process of identifying the future consequences of a current or proposed action.
>
> THE INTERNATIONAL ASSOCIATION OF IMPACT
> ASSESSMENT

The underlying reason for environmental impact assessments is the desire to identify, evaluate and predict the physical, chemical, biological, social and economic effects of industrial and development activities on the existing environment. The statutory and regulatory basis for environmental impact assessments began in 1969 with passage of the National Environmental Policy Act (NEPA) in the United States. In succeeding decades, the concept has become incorporated into laws and practice in more than 200 countries. All such assessments follow the same general principles, with minor differences that can easily be accommodated by process adjustments.

The International Association of Impact Assessment (IAIA) has published "Principles of Environmental Impact Assessment Best Practice" that sums up all the important considerations. These principles are defined below.

2.1 Definition of EIA

Environmental impact assessmentis the process of identifying, predicting, evaluating, and mitigating the biophysical, social, and other relevant effects of development proposals prior to major decisions being taken and commitments made.

2.2 Objectives of EIA

Regardless of geographic location, type of project being evaluated, and statutory authority, an environmental impact assessment has four general objectives—

- To ensure that environmental considerations are explicitly addressed and incorporated into the development decision-making process.
- To anticipate and avoid, minimize or offset the adverse significant biophysical, social and other relevant effects of development proposals.
- To protect the productivity and capacity of natural systems and the ecological processes which maintain their functions.
- To promote development that is sustainable and optimizes resource use and management opportunities.

2.3 EIA Principles

Two levels of EIA principles have been defined by the IAIA:

1. Basic principles apply to all stages of EIA; they also apply to strategic environmental assessment (SEA) of policies, plans and programs. The list of basic principles should be applied as a single package, recognizing that those included are interdependent and, in some cases, conflicting (e.g., rigor and efficiency). A balanced approach is critical when applying the principles to ensure that environmental impact assessment achieve their purpose and are carried out to internationally accepted standards. Environmental impact assessments thus produce both complete analyses and the means of reconciling apparently conflicting principles.
2. Operating principles describe how the basic principles should be applied to the main steps and specific activities of the environmental impact assessment process e.g., screening; scoping; identification of impacts; assessment of alternatives. It is also envisaged

that subsequent levels of principles could evolve e.g., "activity-specific," "state-of-the-art," and "next generation". However, their development would constitute a separate effort, building on and extending the IAIA-defined principles presented below.

2.3.1 Basic Principles

An environmental impact assessment should be—

Purposeful—the process should inform decision-making and result in appropriate levels of environmental and community well-being.

Rigorous—the process should apply "best practicable" science, employing methods and techniques appropriate to address the problems being investigated.

Practical—the process should result in information and outputs which, with problem solving, are acceptable to and able to be implemented by proponents.

Relevant—the process should produce sufficient, reliable, and usable information for development planning and decision-making.

Cost-effective–t-he process should achieve the objectives of the EIA within the limits of available information, time, resources, and methods.

Efficient—the process should impose the minimum cost burdens in terms of time and finance on proponents and participants consistent with meeting accepted requirements and objectives of EIA.

Focused—the process should concentrate on significant environmental effects and key issues–i.e., the matters that need to be taken into account in making decisions.

Adaptive—the process should be adjusted to the realities, issues and circumstances of the proposals under review without compromising the integrity of the process. It should be iterative by incorporating lessons learned throughout the proposal's life cycle.

Participative—the process should provide appropriate opportunities to inform and involve the interested and affected publics, and their inputs and concerns should be addressed explicitly in the documentation and decision-making.

Interdisciplinary—the process should ensure that the appropriate techniques and experts in the relevant biophysical and socioeconomic disciplines are employed, including use of traditional knowledge as relevant.

Credible—the process should be carried out with professionalism, rigor, fairness, objectivity, impartiality, and balance, and be subject to independent checks and verification.

Integrated—the process should address the interrelationships of social, economic and biophysical aspects.

Transparent—the process should have clear, easily understood requirements for EIA content, ensure public access to information, identify the factors that are to be taken into account in decision-making, and acknowledge limitations and difficulties.

Systematic—the process should result in full consideration of all relevant information on the affected environment, of proposed alternatives and their impacts, and of the measures necessary to monitor and investigate residual effects.

2.3.2 Operating Principles

The EIA process should be applied—

- As early as possible in decision-making and throughout the life cycle of the proposed activity.
- To all development proposals that may cause potentially significant effects.
- To biophysical impacts and relevant socioeconomic factors, including health, culture, gender, lifestyle, age, and cumulative effects consistent with the concepts and principles of sustainable development.
- To provide for the involvement and input of communities and industries affected by a proposal, as well as the interested public.
- In accordance with internationally agreed measures and activities.

Specifically the EIA process should provide for—

Screening—to determine whether a proposal should be subject to EIA and, if so, at what level of detail.

Preparation of environmental impact statement (EIS) or report (EIR)—to document clearly and impartially the impacts of the proposal, the proposed measures for mitigation, the significance of effects, and the concerns of the interested public and the communities affected by the proposal.

Scoping—to identify the issues and impacts that are likely to be important and to establish terms of reference for EIA.

Examination of alternatives—to establish the preferred or most environmentally sound and benign option for achieving proposal objectives.

Review of the EIS—to determine whether the report meets its terms of reference, provides a satisfactory assessment of the proposal(s) and contains the information required for decision-making.

Impact analysis —to identify and predict the likely environmental, social and other effects of the proposal.

Decision-making—to approve or reject the proposal and to establish the terms and conditions for its implementation.

Mitigation and impact management—to establish the measures that are necessary to avoid, minimize, or offset (mitigate) predicted adverse impacts and, where appropriate, to incorporate these into an environmental management plan or system.

Follow-up—to ensure that the terms and condition of approval are met; to monitor the impacts of development and the effectiveness of mitigation measures; to strengthen future EIA applications and mitigation measures; and, where required, to undertake environmental audit and process evaluation to optimize environmental management.[1]

Evaluation of significance—to determine the relative importance and acceptability of residual impacts (i.e., impacts that cannot be mitigated).

2.4 Other Guidelines

The "E7 Network of Expertise for the Global Environment" (Électricité de France; ENEL, Italy; Hydro-Quebéq, Canada; Kansai Electric Power Company, Japan; Ontario Hydro, Canada; RWE Energie AG, Germany; Southern California Edison, U.S.A.; Tokyo Electric Power Company,

---

[1] It is desirable, whenever possible, to have monitoring, evaluation, and management plan indicators designed so they also contribute to local, national, and global monitoring of the state of the environment and sustainable development.

Japan) prepared a manual for developing and eastern European countries to follow as they develop electrical power systems. They summarize an EIA as both a planning process and a decision-making process.

The planning process is used to help ensure that environmental concerns are taken into account early in the project-planning process, along with the traditional technical and economic considerations. The process identifies, predicts, interprets, and communicates potential impact information. The decision-making process examines alternative ways of carrying out a project. The EIA provides a framework for gathering and documenting public and external knowledge and opinion. These two processes allow for informed decisions.

In 1977, a team at Virginia Polytechnic Institute (U.S.A.) released a report titled, "A Computerized Method for Abstracting and Evaluating Environmental Impact Statements" [18]. This report focuses specifically on dams and the impacts they create both up- and down-river from the project site. They use the Leopold Matrix method for impact assessment (discussed in Chapter 6 on page 39) and offer "... a method by which the value judgments and estimates of evaluators may be dealt with statistically".

One reason that there is an extensive literature base on the goals and methods of environmental impact assessments is because environmental impact assessments are based on values, concepts, and expressions that cannot be measured using conventional tools.

As will be explained in the rest of this book, the components of the overall process include an *environmental description and assessment* of the current conditions, a description of project *alternatives*, predictions of the *impacts* each alternative might have on each identified segment of the baseline environment, and an *evaluation of impact significance* of the different project alternatives on the baseline environment. The final results are issued in a Record of Decision.

2.5 Problem Areas

Within the EIA process a major cause of problems is the concept of *significance*. The word is commonly found in environmental laws and regulations. Sometimes significance is defined and sometimes it is not. When it is defined, the only certainty is that some stakeholders like the definition and others do not. Everything in the EIA is dependent on significance—it is found in every stage—but it cannot be directly measured. This is a major source of controversy and dissension with the conduct of environmental impact assessments. An in-depth examination of several definitions of *significance* is in Section 6.3 on page 45. Other vague, imprecise concepts (similar to significance and often found in EIAs) include the concepts of *acceptable, sustainable, cumulative,* and *quality* (as in wildlife habitat quality).

There are differences in meaning and understanding between concepts we can directly measure and those we cannot. The latter terms are critical to communications and generalization. We can measure *distance,* but we cannot measure *near* or *far*. We can measure *size* but we cannot measure *small* or *large*. If we limit our thinking and communications to measurable entities, and we use Boolean logic and probabilities to include or exclude entities and events by discontinuities or thresholds, we miss the richness and variety inherent in the natural world. To capture this richness and variety we need a way of expressing values— concepts such as *significance, quality, relative distances,* and *comparative sizes*. This method of expression is the use of *linguistic variables*. To understand what is meant by "linguistic variables" consider these two quotations by Lotfi Zadeh and Bertrand Russell—

> In retreating from precision in the face of overpowering complexity, it is natural to explore the use of what might be called *linguistic* variables, that is, variables whose values are not numbers but words or sentences in a natural or artificial language.
>
> The motivation for the use of words or sentences rather than numbers is that linguistic characterizations are, in general, less specific than numerical ones[43].

> All traditional logic habitually assumes that precise symbols are being employed. It is therefore not applicable to this terrestrial life but only to an imagined celestial existence[27].

As Zimmerman [45] notes (on his page 141), Zadeh's "quotation presents in a nutshell the motivation and justification for fuzzy logic and approximate reasoning." It is the use of these computational intelligence techniques that permit the resolution of the problems that may develop when planning and conducting an environmental impact assessment.

To illustrate the importance of linguistic variables to the timely and cost-effective completion of environmental impact assessments, consider the term *steep*.

There are many ways to measure the angle of a slope to very high degrees of accuracy. However, there is no way to measure steepness. That is an inherently imprecise term that means different things in different contexts. For a railroad, a track grade greater than 2 percent is too steep for freight trains while passenger trains can generally climb a 4 percent grade. In a mountainous area, a 100 percent grade might not be considered to be too steep to climb by foot. The problem with using a term such as *steep* in a regulatory environment (such as prohibiting development on "steep slopes") is that any such threshold between steep and not steep is arbitrary and meaningless. If, for example, the threshold is set at 20 degrees, does this mean that a measured slope of 19.5 degrees is not steep and could be developed? Probably not. So arguing over whether a particular measured grade is steep is neither productive nor does it address the underlying problem. How a linguistic variable (such as steep) is related to the measurable value (percent slope or grade) is explained in detail in Chapter 9 on page 63.

2.6 EIA Process Overview

The entire environmental impact assessment process is concerned with balancing societal values and managing environmental risk. Each step identifies those values, determines where conflicts exist or decides whether the project is consistent with those values. If the decision is to permit the project then the process is used to determine constraints and mitigation appropriate to manage and minimize environmental risk. Science, economics, and sociology are important components, but the values and risks are expressed using terms that are inherently imprecise, vague, unmeasurable, or fuzzy. That is, they are expressed as linguistic concepts or linguistic variables.

Using fuzzy terms are the way most people think of environmental conditions and impacts. Emotional feelings and values cannot be measured but are meaningful and valid in expressing beliefs. Trying to accommodate these fuzzy concepts in environmental impact assessments has not been successful—at least, not in consistently meaningful and supportable ways. Add to this the different feelings and values held by different groups in society and the problems multiply. Perhaps this explains why some environmental impact assessments leave the determination of scope and significance to the "experts" in the lead agency and related organizations. The experts speak the same language, hold the same (or very similar) values, and can usually reach agreement on what they believe is important. However, the stakeholders left out of the process frequently feel slighted and angry.

While many non"experts" among the public and other stakeholders do not have specific knowledge and experience equal to that of agency staffers, their values and opinions are crucial to the success of an environmental impact assessment. This makes sense if we think of the reason for conducting the assessment in the first place: to meet societal standards for acceptable development and industrial activities. Sometimes, concerns raised are red herrings that attempt to camouflage the underlying NIMBY[2], or they are based on ignorance of the dynamic nature of ecosystems. Regardless, the concerns and values must be incorporated into the process if for no other reason than to reduce the risk of legal challenge and increase general acceptance.

There are a number of components which may or may not be included in an environmental impact assessment of a specific project. The challenge is finding methods and techniques that identify impacts, predict future environmental conditions, evaluate significance, and reflect societal values. The general goal is to effectively manage environmental risk, and such management can only be effective when the information is suitable to the task and the decision-making process is as objective and comprehensive as possible.

The entire process begins by screening the proposed project to determine whether or not it might be expected to have significant negative impacts on the existing environment. This screening is usually set by statute or administrative regulation and may be based on a checklist of physical (geologic and geographic), chemical, biological, ecological, economic, social, cultural, and infrastructural components. The lead agency uses this information to decide whether it is likely that the project, as proposed, is going to have meaningful impacts on various stakeholders and the environment.

Screening is supposed to be objective, quantitative and straightforward. This is inherently impossible for many reasons, e.g., the sub-

---

[2] Not *in my backyard!*

jectiveness of *significance*, the inherent variability of natural ecosystems, the lack of sufficient data to characterize the environment, and our inability to readily distinguish important components from less important ones. Also, it is difficult to characterize something as complex as an ecosystem in a single number that is both meaningful and based on sound and reasonable assumptions. Diversity indices were once thought to be such a useful characterization, but in practice they fail to be practical and useful.

When the screening concludes that there is a likelihood that the project will have significant environmental impact the process moves to the next process: scoping.

Scoping determines what components are to be included in the EIA and what alternatives are to be considered. Scoping usually occurs in several phases. First, the project proponent (the applicant, if a permit is required) identifies the preferred alternative and usually comes up with at least two others: the "No Action" and "Worst Case" alternatives. Then the regulatory and resource agencies add their input and, in the last phase, input from other stakeholders (e.g., supporters, opponents, neighbors) and others are solicited. It is during the scoping process that all stakeholders have an opportunity to express their values on which components should be included in the assessment and their opinions on what additional alternatives should be considered. Manual (and subjective) processing of alternatives can become onerous financially and for the amount of time consumed. A modern, objective process requires very little additional time and cost to add alternatives. However, there will almost certainly be some initial decisions by the regulators with regard to which alternatives to include so as to represent a reasonable range. In the United States five to seven alternatives are usually considered adequate. With such a broad representation of interests participating in the scoping process, conflicting agendas can appear. The agenda of the project proponent and supporters is obvious: they have determined that there is an economic gain to them from the development and operation of the project. Or, in the case of a government-sponsored project (for example, a flood control dam on a river), there is a public need to be met by completing the project. All too often, there are opponents of the project. They might have an agenda of not wanting any development at that location or they may be a business competitor whose agenda is to obtain the rights to develop and operate the project themselves. The regulatory agency or other decision-maker may be caught in the middle of these conflicting agendas, or they may have their own agenda to promote in the process. These conflicting agenda are accommodated in the scoping process described in Part II of this book.

The public input sessions and the decisions of what components and alternatives to include in the EIA can take a comparatively long time, be contentious, and delay the project significantly. However, once the project is scoped it moves to the technical stages.

The three, technical stages of environmental impact assessments are environmental inventory of the baseline condition,; impact evaluation (prediction of change, estimate of change magnitude, measure of change significance) of the defined alternatives, and the written report (variously called an Environmental Assessment, Environmental Impact Statement or Environmental Impact Report).

It seems obvious that no assessment can be made without comparison to a standard. In the case of an EIA for industry and development, that standard is the existing environment. Yet while there are published guidelines on what components to include in the baseline and how to collect the data, methods of characterizing the baseline conditions are uncommon. Almost always, the individual components (e.g., wildlife habitats, hydrology, air quality) are evaluated for change under each alternative. Characterization of the entire set of conditions is critical so that all alternatives—including the "No Action" alternative—can have their impacts compared against the same standard. Using words alone is insufficient. There is the possibility of picking and choosing only those aspects of the existing environment that support a particular, desired outcome in the alternatives analyses.This oversight will be corrected by the approach presented in Part II.

Other reasons for describing the baseline conditions include identifying the presence of species of special concern (such as a listing, or being a candidate for listing, under the Endangered Species Act [ESA]) and already degraded components of the local ecosystem. As examples, a stream reach may have poor water quality from agricultural runoff, a section of range may be overgrazed from years of poor grazing practices, or the land might have supported a timber treatment plant which left the soils and subsoils contaminated with arsenides and organic chemicals (i.e., what the US Environmental Protection Agency calls *brownfield* sites). Recognizing opportunities for improvement of existing environmental conditions by the proposed project can only be done when the area is quantitatively and objectively characterized.

The impact evaluation stage of the EIA process is where everything comes together. It is in this stage that impacts are identified, their effects forecast and their significance within both local and broader contexts evaluated. There are multiple ways in which to predict future environments based on identified impacts and even more ways of determining whether or not they are significant. The recent review by [26] presents eight different definitions or interpretations of *significance* as published between 1984 and 1999. Despite the core value of significance to the evaluation of impacts, there is no consensus definition of the term. Because of this imprecision in the critical decision factor, controversy and legal challenge is an almost inevitable result.

Preparing and presenting the EIA report is important for two reasons. First, it must effectively communicate highly technical components and subjective values to the nontechnical decision-makers. Second, it must clearly and compellingly document how data were evaluated, interpreted and used to reach a recommendation (or decision). Most of the time there is a prescribed format for the report by the regulating authority but the quality of the writing is an important consideration. The report must answer the questions, "so what?" "why?" and "how?" so that every reader understands. While report preparation is not as subjective and controversial as are other stages in the EIA process, it is the public and official face by which the entire process is judged (sometimes, literally). Therefore, in order to make this book complete, this topic will be covered in Chapter 7 on page 53.

3

Scoping

Determining the scope of the EIA must be the first formal stage in the process. It is the first opportunity for public involvement and can set the tone of the developer—public interaction. Considering the potential importance of scoping in successfully completing the EIA, the limited number of publications available is surprising. What publications do exist focus more on public participation during the environmental impact assessment process than they do on the importance of determining components and alternatives to include in the assessment. An example of the need for public participation is found in [14]; they devote a chapter to public participation in the UK and the European Union. More comprehensively, the Canadian Environmental Protection Agency (CEPA) demonstrates the importance of public participation in the scoping process. As described by [17], the CEPA includes Aboriginal peoples' values and needs in the EIA process. Canter [4] includes a chapter discussing public involvement in the overall EIA process. The chapter considers both the advantages and the disadvantages of involving the public in the scoping project. Canter also includes a model for such participation. Abundant references can be found that describe the scoping process and results in specific environmental impact assessments, but there are very few texts or other academic considerations of the rationale, process, and goals of scoping. From the regulatory perspective, both France and the U.S. require public involvement in scoping when the project is of sufficient size or potential adverse impacts. Otherwise, such participation may not be mandated by statute or regulation.

The scoping stage serves three functions within the environmental impact assessment process:

1. It is the first opportunity for public input.
2. It identifies environmental, social, economic and cultural components that might be considered in the assessment.
3. It allows for suggestions of more (or different) alternatives than the project proponent might have recognized.

Public input to the assessment process takes different forms depending on the jurisdictional level and project size. Construction of a boat launching facility (ramp and parking lot) on a lake is quite different from construction of a multi-megawatt electrical-generating plant. The former is probably under the permitting authority of a state or local jurisdiction and could be expected to have little opposition and few environmental impacts. The latter is almost certainly under the regulatory control of an authority at the national level and will be closely examined by many different interest groups.

For small projects with limited and local potential impacts (such as the recreational boating facility example) public participation during the scoping phase may consist of a notice published in the local newspaper, a mailing to adjacent property owners, and notification of other state regulatory and resource agencies. Written comments are solicited and the lead agency determines the scope of the assessment.

For major developments with the potential for regional or national adverse impacts (e.g., the power plant example) the scoping phase may involve several public hearings, widely publicized to bring in as many stakeholders as possible. At the hearing the project is explained, comments are elicited, and the process repeated at different locations and dates. While the lead agency still makes the final decision on the scope of the assessment process, attendees of the meetings should be kept informed of the decisions.

With regard to the components addressed during the assessment, there are generally accepted lists of physical, chemical, biological, social, cultural, and economic factors for all environmental impact assessments. Quite often the scope of the components to be considered in the EIA is decided by the lead agency or the technical team preparing the assessment and report. Other projects have a major effort to elicit public participation in determining what factors should be included in the assessment. In any case, no matter how extensive the list it needs to be adapted to the specific project and location and only the most important components should be considered. The latter issue is where much disagreement arises.

Local residents frequently have more intimate knowledge of the local ecosystems than does anyone from other areas. The insight can be very helpful in ensuring that all factors are at least considered at the front end of the project and not after much time and expense has been spent on the assessment. In a similar vein, people not deeply familiar with planning for the project may suggest alternatives that had not occurred to the project team or the lead agency. Practical or not, the input can be very useful. Just because an issue or alternative is raised during the public scoping process does not mean that it is, or should be, automatically included in the assessment. The rationale for inclusion or exclusion must be clearly and convincingly presented. Under many existing assessment methods that have been used, this is frequently very difficult to do effectively.

All too frequently there are well-organized objectors to a project. These objectors will use their exclusion from the scoping stage as a basis for administrative or legal challenge in an attempt to delay the project and drive up the costs before there is any economic return to the developer. Including the objectors' opinions at this very early stage of the process removes the complaint that no one listened to their views or gave them sufficient consideration. This factor alone can justify the comparatively minor costs of soliciting *all* public input at this stage. While the traditional approach of conducting an EIA does not specify a particular method for incorporating and evaluating those comments, the modern approach does.

Overall, the advantages of public participation in the scoping process include an opportunity for all interested groups to express their opinions, for the public to see that the responsible agencies listen to their concerns and that the process is open and as objective and inclusive as possible. The disadvantages of public participation include having erroneous information presented by those who strongly oppose the project, creating confusion by the introduction of new components or issues, the potential for delay in the project and increased costs.

Public involvement—particularly during the scoping process—fails unless the information received by the lead agency and project team is incorporated into the EIA process. In particular, public feedback helps better to define project need as well as identify unique or subtle features of the baseline environmental conditions and potential (or actual) environmental impacts. The challenge has been to incorporate public input in a consistent and meaningful way.

Without formal methods for classifying, evaluating, and deciding on what concerns, components, and alternatives will be considered in the assessment, the process is unwieldy and overwhelming. In some cases, spreadsheets have been prepared and manual classification and evaluation have been done by teams of attorneys and paralegals. This situation can be greatly approved by both changes in how the scoping is conducted and by how the issues and concerns are processed and analyzed.

In Part II an effective procedure to process subjective values, issues, and components to be incorporated into the environmental impact assessment is explained in detail.

4

Baseline Conditions

Baseline conditions may also be referred to as the *environmental setting*, *existing conditions*, and other similar terms. The baseline conditions are the physical, chemical, biological, social, economic, and cultural setting in which the proposed project is to be located, and where local impacts (both positive and negative) might be expected to occur.

These conditions are the standard against which are compared projected future conditions from project alternatives. Their description and characterization are necessary for decision-makers, reviewers, and others who are unfamiliar with the project site and surrounding landscape.

Unfortunately, there are few published directions or guidelines on how to apply the descriptions of the baseline conditions. To understand why directions or guidelines are necessary requires examination of what roles the baseline conditions play in the EIA process.

Every impact assessment is (or, at least, should be) conducted with reference to a standard: the current environment of the area in which the proposed project is to be located. The baseline conditions usually includes components in the broad categories of physical-chemical, biological, cultural, and socioeconomic factors. Considerations in describing the baseline conditions are:

1. What components are to be included or excluded.
2. How the necessary data are to be collected and analyzed.
3. How the baseline conditions can be objectively compared with future conditions under alternative scenarios.

4.1 What To Include

Many, if not most, regulatory agencies have a list of environmental
components to be considered in an environmental impact assessment.
In the western United States the initial list of components is usually
based on the type of environment in which the specific agency reg-
ulates development and by the type of project proposed. The components are usually assembled into a checklist based on statutory requirements, administrative regulations, public scoping or the anticipated impacts of the project. A representative example of an existing
approach is the list used by the Battelle Method for water resource development projects ([10]), which includes—

ECOLOGY
Terrestrial Species and Populations

- Browsers and grazers
- Crops
- Natural vegetation
- Pest species
- Upland game birds

Aquatic Species and Populations

- Commercial fisheries
- Natural vegetation
- Pest species
- Sport fish
- Waterfowl

*Terrestrial Habitats and Communi-
ties*

- Food web index

AESTHETICS
Land

- Geological surface material
- Relief and topographic charac-
ter

- Width and alignment

Air

- Odor and visual
- Sounds

Water

- Appearance of water
- Land and water interface
- Odor and floating material
- Water surface area
- Wooded and geologic shore-
line

PHYSICAL/CHEMICAL
Water Quality

- Basin hydrologic loss
- Biochemical oxygen demand
- Dissolved oxygen
- Fecal coliforms
- Inorganic carbon
- Inorganic nitrogen
- Inorganic phosphate
- Pesticides

- pH
- Stream flow variation
- Temperature
- Total dissolved solids
- Toxic substances
- Turbidity

HUMAN INTEREST/SOCIAL
Education/Scientific

- Archaeological

- Ecological
- Geological
- Hydrological

Historic

- Architecture and styles
- Events
- Persons
- Religions and cultures
- Western frontier

Other checklists are different in content, can easily contain 50—100 items, and usually reflect the type of project being assessed. It is important that components deemed significant and likely to be affected (positively or negatively) be selected for examination. Because *deemed significant* varies with the individual or interest group it is a subjective evaluation and is a potential source of challenge and conflict. This potential is one of the concerns that is resolved using the modern approach described in the next part of this book.

4.2 Collecting Data

Ecosystems are highly dynamic. They vary both spatially and temporally, and the rate of variation can also change. This multidimensional variability is very important because it places severe restrictions on the quantity and quality of data that can be used to characterize the baseline conditions in the project area. Consider, as examples, collecting species-level information on plants and aquatic insects.

In the case of the plants, it is important to do the field work when the plants are in bloom (or otherwise unambiguously identifiable), and it is very important to consider the seral (successional) stage of the vegetation.[1] This knowledge is applied later when the "No Action"

---

[1] All plant species assemblages (communities, in the vernacular) undergo change over time. Bare ground, or heavily disturbed ground, is first colonized by annual plants. After a few years perennial plant species predominate. Vertical layers, too, change over time. The first plants to invade are grasses and other ground covers. Eventually forbs and shrubs colonize the

alternative is considered and the future of the otherwise unmodified environment is predicted.

In the case of aquatic insects and other aquatic biota both time and place of collection are very important factors for interpreting the data. In temperate climates the majority of the insect species are present during winter (when it is less comfortable to be collecting them) and fish or aquatic plant species may or may not be present at specific times of the year. There is also a stronger spatial affinity for particular habitats and locations by aquatic biota than is seen for many terrestrial biota.

Because there is never enough time and money to describe comprehensively the inherent variability of the baseline conditions, it is necessary to use a variety of sources for baseline data. These data represent a snapshot of particular times and locations.

4.2.1 Literature Reports

This category includes academic and agency studies of the project area as well as other published data from previous projects or proposals in the vicinity. Many federal and state agencies have digital databases on physical, chemical, and biological components of the area. County and city departments may have pertinent information on cultural, historic, economic, and infrastructural factors that can be used to describe the baseline conditions. When incorporating these data with those from other sources it is necessary to report the dates of collection in order to properly interpret them.

4.2.2 Field Studies

There is always a need to collect current data on the baseline conditions from the site. Endangered species surveys, wetland delineations, and condition of the vegetation and ground are required and usually not sufficiently recent from prior work. Sometimes methods to be used are specified by the regulatory or resource agencies; at other times the methods are those widely accepted within the scientific community. The methods that are the most appropriate and useful should be used.

---

area, and they are eventually replaced by trees in many environments. Even the so-called *old-growth* or *climax* stages are temporary, but they tend to last longer than the preceding stages.

The attention given to choosing such methods should be as rigorous as necessary.

The extent and amount of detail for all project-specific field work needs to be related to the data use. If a component will most likely not be affected by the project, or if that component is not deemed significant (by the scoping process results), then less effort should be spent on data collection for that component. For those components that are certain, or likely, to be affected by the project the effort expended on characterization needs to be appropriate to the analytical and interpretive methods and the attributes that the project might affect.

A factor not frequently enough considered when establishing baseline conditions is the broader landscape in which the proposed project is located. If extensive fish and wildlife habitats are available regionally, their on-site value may be quite different than when they are few or rare at the larger spatial scale. For example, a location in a temperate, humid forest could represent a very small fraction of all the available habitats suitable for an assemblage of plant or animal species. On the other hand, a location adjacent to a small water body in a semi-arid, shrub-steppe landscape might be the only aquatic habitat and water source for many kilometers in any direction. Therefore, the importance of the habitats at the latter site are much greater than those of the former site.

The importance of including the project site's relation to the landscape is that many included components are more extensive than the project site itself. The temporary or permanent loss of a population of a particular species of plant differs in significance if the location is the only one known in which that species is found compared with a location well within the known distributional range of the species or a location near the edge of the species' distribution. Ecologically, each of these three situations is distinct. If the baseline conditions does not include the description of the landscape then it is not possible for decision-makers, or other interested parties, to be adequately informed.

4.3 Baseline Condition Use

The description of the baseline environment is usually a report augmented by tables, figures and extensive technical appendices detailing how the data were acquired. The way the description generally is used to assess impacts by the various alternatives is component-by-component. This is contrary to the concept of ecosystems, where components are integrated and act both collectively and on each other, not in isolation. Socioeconomic systems are also integrated and their dynamics cannot be understood by examining individual components out of context with other components.

There are two serious deficiencies with the way most conventional reports present baseline conditions:

1. They do not permit quantitative comparisons of predicted future conditions produced by each alternative to the current conditions.
2. They do not reflect the value (or "significance") of the area in relation to the broader landscape or the values of various stakeholder and other interest groups.

These deficiencies open opportunities for challenge to what was done, how it was done and how the final decision was reached. Even the casual reader of this information may have difficulty understanding its meaning. The textual description of the baseline conditions has no expression of significance. That is, there is no defined scale of environmental condition against which the current values can be compared. These two deficiencies need to be removed, and they are in the quantitative approach described in Chapter 10 on page 115.

4.4 Missing or Insufficient Data

When characterizing baseline conditions, missing or insufficient data are common because ecosystems are dynamic and highly complex. For example, consider a tract of recently disturbed ground (after a range or forest fire or a volcanic eruption such as that of Mt. St. Helens in Washington state on May 18, 1980). A survey of the vegetative communities may find no living plants. Depending on the time of year, it may be two to nine months before the pioneer species appear (primarily grasses). The following year, perhaps in two years, the grass species have changed to some extent and forbs and shrubs are appearing. Even five to ten years later the vegetative community is variable and subject to change. Under these conditions—and all other situations just like them—the definition of "complete" data is a judgment call. In addition, there is never adequate time and money to collect baseline data long enough to establish the inherent, natural variation in the ecosystems.

On the less technical side, there is a real problem with defining "missing or insufficient data." There is no standard that defines how much data are sufficient to characterize baseline conditions at a site. Missing and nonexistent data may not be the same. These are not abstract, philosophical discussions; they need to be addressed on a project-by-project basis to allow useful data collection. When the scoping process has been competently carried out, the project proponents know the concerns that the EIA must address. Working back from those concerns eventually leads to the types and amounts of data that need to be assembled about the environmental components in and around the project location. However, even when the environmental components are well defined, and the data to characterize them are established, it may not be practical to spend years measuring these components completely. It would be valuable to be able to use qualitative data to supplement measured data and to replace missing or incomplete data. More than merely "valuable," qualitative data can be crucial to developing a sufficient basis for supporting a technically sound and legally defensible analysis.

Fortunately, there is a mathematically (and ecologically) sound method to overcome most, if not all, of these concerns about missing or incomplete data. The technique of mixing measured values with qualitative evaluations by observers is described in Section 15.5 on page 213. This method will fill the gaps where detailed measurements and surveys cannot be undertaken because of areal extent, time constraints, or cost considerations.

5

Alternatives

It is surprising how little guidance is offered to EIA preparers on the description of alternatives. Two books listed in the references (i.e., [4] and [14]) do not address the subject. A search of the world wide web finds little in the way of explicit direction. One useful statement by the US Department of Interior offers, "[b]ased on the information received during the initial scoping effort and other information, such as the location of sensitive natural resources, ... we identify alternatives to the proposal that might reduce possible impacts." This helps to identify alternatives, but offers nothing on the structure of their description.

According to the US Council on Environmental Quality (CEQ), alternatives should be described according to these guidelines—

> This section is the heart of the environmental impact statement. Based on the information and analysis presented in the sections on the Affected Environment[1] ...and the Environmental Consequences ..., it should present the environmental impacts of the proposal and the alternatives in comparative form, thus sharply defining the issues and *providing a clear basis for choice* [emphasis added] among options by the decision-maker and the public[7].

There is almost always a question of how many alternatives need to be presented and analyzed. Under the traditional approach, there may

---
[1] That is, the existing conditions.

be greatly increased costs associated with the development and characterization of each alternative. However, it is appropriate to explore rigorously and evaluate objectively all *reasonable* alternatives. That is, those alternatives that can be implemented at a positive benefit:cost ratio for the developer. For alternatives which were eliminated from detailed study, the reasons for their having been eliminated should be documented.

While the CEQ suggests that the assessment "... [d]evote substantial treatment to each alternative considered in detail including the proposed action so that reviewers may evaluate their comparative merits." One of the most common reasons for the rejection of a draft EIS in the US is that the alternatives are not equally considered. The preferred alternative has the most extensive description, while the "No Action" and "Worst-Case" alternatives are given less attention. When the discussion and evaluation of alternatives are limited to subjective methods, this uneven treatment is to be expected. If objective characterization of all the alternatives can be done, then there is no longer any reason to avoid "substantial" treatment of each alternative in the environmental impact assessment.

The "No Action" alternative is usually insufficiently addressed. To some, the denial of project permits means that the existing conditions remain untouched and unchanging. However, ecosystems are not static, they are effected by natural and anthropogenic effects beyond their borders and the existing conditions may not be desirable by today's societal values.

All natural systems evolve. Plant assemblages change over time from ground cover to mature forest. Wetlands and bogs may fill in and become terrestrial habitats. As the vegetative types change, so do the types of animals that are found there, and the uses they make of those plants. Rivers meander across their valley floors and can change from a single channel to a braided one, and back again. Sharp meanders are cut off during high flow events and the isolated former loops become oxbow lakes. Volcanic eruptions decimate the landscape, but plants and animals begin to return during the next growing season and the cycle begins again.

Humans have changed environments for as long as we have been around as a species. There are fossil records of mammoth bones at the bases of ancient cliffs. A few days' food was apparently obtained by driving a herd over a cliff when not all the meat could have been eaten before it spoiled. Prairies in the Great Plains of the United States used to be periodically burned for various human-related needs. The deserts of the Middle East were once forested, until fuel and housing needs extracted material more rapidly than it could grow and be replaced. On the beneficial side, in the mid-1800s as the western United States was colonized by migrants from the east coast, the tall sagebrush (*Artemenesia* species) was cleared by sod-busters for farming and grazing of cattle and sheep. When sagebrush regrew it was with varieties closer to the ground. The sage grouse (*Centrocercus* species) populations increased because they could now reach the leaves and seeds of the plants that, before, had been too high for them. Acid precipitation in the northeastern United States was reported to have originated from the high-sulfur content exhausts of coal-fired power plants in the midwest. Radiation from the Chernobyl nuclear plant accident has been detected globally.

Many areas in the world have been degraded by war, poverty, political indifference, and ignorance. These areas, and even more that do not appear as obviously disturbed, can benefit from planned modification and improvement to bring them into alignment with current societal standards. Quite often, only the private sector has the money to effect such changes as mitigation for project effects, as the result of designed reclamation or as an amenity that increases the economic value of the land and the associated development.

The "No Action" alternative should discuss all these actual and potential factors when determining the positive and negative impacts of not allowing the project and projecting how the future of the site will appear.

The "Preferred Alternative" is the one around which the entire project concept was developed. Depending on whether the project initiation is in the private sector or the public sector of the local economy, the motive for developing the project may be profit, reduced costs or other similar vested interests. The preferred alternative usually evolves from internal discussions and developer needs so it tends to maximize the desired outcome. When project approval must come from higher in the corporate, bureaucratic, or political hierarchy, the benefits are stressed while the costs are minimized. This is a very reasonable approach for those who need to "sell" the project to the internal decision-maker. As a result of the internal planning and organizational processing, the preferred alternative is generally defined to a high level of detail with a strong emphasis on the positive aspects. It is implicitly assumed that the project will be approved by regulatory authorities when planning is done under the preferred alternative.

When the preferred alternative is described for inclusion in an environmental impact assessment, the potential impacts are identified and the standard hierarchy of avoid, minimize, or mitigate is applied. It is very reasonable for the project developer to want to achieve the minimal level of mitigation required so as to minimize project costs. There is absolutely nothing wrong with this approach. If the mitigation can be shown to achieve the desired goal then the least expensive way is the one that should be used.

There are occasions when the appropriate mitigation measures are not already included in the proposed action or alternatives. This may happen when the project team that develops the environmental impact assessment does not have adequate expertise available. If this occurs, and is noticed, then these other mitigation measures should be added to the document for analysis.

Alternatives, while representing different options for accomplishing the project goals with different costs, profitability, and environmental affects, need to be presented in a way that permits each alternative to be quantified. The value of each alternative allows it to be compared among others as well as to be applied to the baseline conditions. In addition, each environmental component considered to be significant must be addressed in the quantified description of each alternative.

It is possible to measure with a known degree of accuracy the effect of a project alternative on some environmental components. For example, it is easy to note the amount of permeable surface that will be paved and made impermeable, or the area of wetlands that would be filled. Other components of the baseline environment cannot be easily counted (e.g., the number of nesting raptors that might be displaced) but these effects can be expressed in comparative terms such as *some, many*, or *few*.

6

Impact Assessment

This broad subject is the focus of most books and papers on environmental impact assessment. And rightly so, for it is highly complex, extremely difficult to select the "right" method, and too frequently unsatisfactory to one party or another regardless of what method has been adopted. However, it is the impact prediction and evaluation of the various proposed alternatives that form the basis of a decision. Therefore, this is the heart of the entire process and the one that properly has had the most attention.

There are three factors that make up the assessment of a particular alternative on the baseline environment. The first factor is whether the alternative will cause a measurable (or noticeable) change in the current environment. If there will be, or could reasonably be expected to be, such a change, its direction must be noted. The term "impact" is neutral and means change. Change can be positive (beneficial) or negative (harmful). Too many environmental impact assessments assume that all change will be negative and ignore, or do not seek, positive change that can be brought forth by an alternative.

The second factor is the magnitude of change and the length of time over which it is effective. Some project changes are permanent and easily quantified (e.g., the filling of all wetlands on the site and creation of a building footprint and parking lot on the area). Other changes are temporary (the ten-year operational life of a mine) or difficult to predict with a valid measurement (the change in the æsthetics of scenic value). Much effort by many people and groups over the past three decades have been devoted to creating methods to measure change magnitude in a satisfactory way.

The third factor is a measure of significance of the impacts. Some assessment methods do not address this factor at all, while others have proposed various approaches to creating an objective measure from a subjective value judgment. Significance, of course, is critically important to the decision-makers and the basis for whatever decision is made. These values and judgments (whether "best professional" or otherwise) are much more commonly used in environmental impact assessment than are rigorous, repeatable scientific methods and verifiable data. This reality illustrates why the formal (and informal) methods that have been used for this task all fall short, and why a method based on mathematically rigorous quantification and manipulation of these values is needed.

6.1 Impact Identification

Rather than taking a component-by-component approach that considers what affects an alternative has on air quality, water quality, wildlife habitats, and other components of the existing environment, an overall framework should be used. The appropriate framework provides consistency and ensures that the impact analyses are conducted from the perspective of a complete ecosystem and not as isolated factors that do not interact.

Over the years many approaches have been proposed to fill the need for a comprehensive framework that permits the identification of impacts on the baseline conditions by the different project alternatives. Each proposed solution has advantages and disadvantages that should be considered in light of the specific assessment being conducted.

Ad hoc **approaches** These are usually templates designed for a specific type of project (e.g., a hydroelectric dam) or for a specific geographic area (e.g., Washington State's *SEPA Checklist*). They may not be the best, or the best suited for a specific project, but they are usually required by the lead agency or regulating authority so it is what everyone uses.

Checklists A list of possible impacts (such as the Battelle method list in Section 4.1 on page 28). Each item is presumed to have equal weight and value in the assessment. A checklist is a good starting point but it is not sufficient by itself.

Matrices   A matrix (the Leopold matrix is the most common) adds a second dimension to a checklist (Figure 6.1). The additional

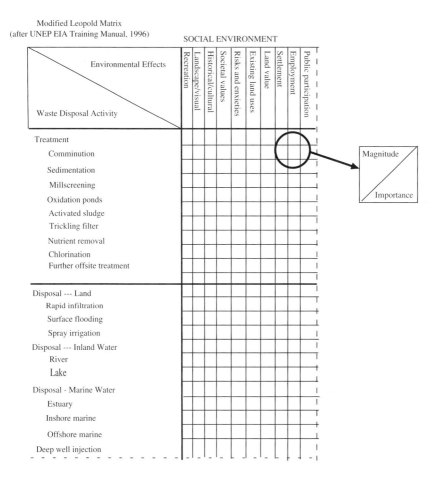

Fig. 6.1. An example of a Leopold matrix. Each cell is scored for impact magnitude and importance.

dimension is a list of project tasks or stages arranged as column headings while the environmental components are the row headings . The cell contents where rows and columns intersect contain information ranging from a simple mark (that component is potentially impacted by an activity) to more complex schemes that score (subjectively) the magnitude of impact and its importance (as represented by a weighting factor). The two drawbacks of matrices are that they do not include values of significance and they leave the decision-makers with the task of interpreting a table of numbers; sometimes a very large table with many numbers.

Networks Networks show the relationships among project needs, specific resulting activities, environmental factors, and other affected interests. While illustrating the complex input–output interactions among an arbitrary number of entities, it does not provide a basis for comparing various alternatives, quantifying pre- and post-activity conditions, assigning and evaluating relative weights or indicating significance. An example of an impact network for dredging is shown in Figure 6.2.

Map Overlays/GIS This is a visual process where a base map is printed on paper and overlays for a particular "theme" (e.g., wetlands, wildlife habitats, land ownership) are printed on a transparent medium such as Mylar®. Inspection shows where two or more "themes" intersect at the same geographic location. The modern tool for this task is the geographic information system (GIS) which can be used simply as an automated mapping tool or can form the basis of spatial analyses (answering questions such as, "What is at this location?" "Where is ...?" or "What is the shortest route between point A and point B that avoids wetlands but is close to roads?"). While the visual presentation is helpful in making the spatial relations of activities and potential impacts easy to grasp, it does not provide a measure of importance, relative weights, or significance.

Models   The GIS is one example of a computerized model used for environmental impact assessments. Other models are numerical (e.g., runoff and erosion), physical or descriptive. The latter is the most common type of impact model. It may reference a checklist or matrix, but the identification

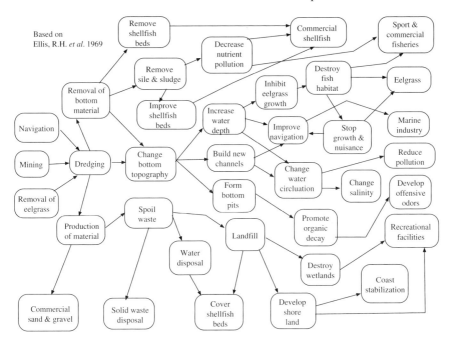

Fig. 6.2. An example of an impact network used on a dredging project.

of impacts is done with words. It is difficult for decision-makers to evaluate different impacts associated with different project alternatives using only a textual description.

6.2 Impact Prediction

Once an impact has been identified more work remains to be done. Naming the impact does not adequately describe the nature of the impact. For example, the impact can be positive (beneficial, such as creation of a storm water pollution control system) or negative (harmful, such as plowing a field up to the bank of a stream). In order to predict the effects of an identified impact a simplified model of natural ecosystem functioning is applied. Many assumptions can be involved during this process. The criteria for characterizing impacts include—

- The impact's nature: whether the impact is positive, negative or neutral; direct or indirect.
- The areal extent of the impact, the volume covered; distribution.
- An estimate of the impact's magnitude (perhaps as "low," "moderate," or "severe").
- The impact's duration: temporary or permanent, intermittent or continuous.
- Whether the impact is reversible or irreversible.
- Whether a negative impact can be mitigated; if so, to what extent.
- The timing: during development/construction, operations or both; immediate or delayed.
- The likelihood of the impact occurring: probable, possible, not known for certain.
- The significance scale: local, regional, national, or global.

The presence of other projects (completed, in development, or planned) will affect the cumulative impact prediction. Like individual project impact predictions, cumulative impacts can be positive or negative as well as offsetting or intensifying. Regardless, because impact predictions are forecasts they need to have a measure of confidence, or certainty, associated with each one. This measure is not a weight of importance or significance but a measure of confidence in the goodness-of-fit of the prediction. Calculating a technically sound confidence measure can be very difficult to do well.

Traditionally, the methods used to predict impact are "best professional judgment," case studies as examples or analogies, quantitative mathematical models, statistical models, pilot (or scale) models, and experiments. With any of these methods, there are three categories of uncertainty associated with the prediction of impacts:

1. Data uncertainties related to limited data or analytical methods whose assumptions cannot be met.
2. Scientific uncertainties associated with inherent variability and insufficient understanding of ecosystem or population dynamics.
3. Policy uncertainties created by imprecisely stated objectives, standards, and regulations as well as subjectively applied concepts and ideas.

6.3 Impact Evaluation

Once an impact has been identified and a prediction made whether it will occur, it is necessary to evaluate the impact's significance. *The evaluation of impacts is not science; instead, it reflects political values and social acceptability.* Despite this subjectivity, impact evaluation for *significance* is the only way that alternatives can be evaluated, future environments compared with the existing environment, and decisions made.

There are two broad aspects to impact evaluation: evaluating each impact with regard to identified values and accumulating impacts for their collective (or cumulative) effects. Impacts (such as a change in the area of habitat for a particular wildlife species, the number of jobs in the area or visibility) are individually evaluated using identified values. These values may include environmental standards (such as those for water and air quality). The "identified values" are those of social acceptability and they reflect the level of public concern. These evaluations are almost always qualitative or relative to threshold values (such as water quality standards). However, the evaluations do have a base in established protocols or guidelines. Once the identified impacts have been individually evaluated it is necessary to combine them in a meaningful way to determine the predicted cumulative effects on the environment. This, too, tends to be done by "best professional judgment" or other subjective criteria. In summary, impact evaluations attempt to assign significance in a rational and defensible way. But, they are still based on subjective values and uncertain measures.

One widely promoted guideline [13] suggests that significance can be tested by answering three questions:

1. Are there residual environmental impacts?
2. If there are residual impacts, are they likely to be significant?
3. If the residual impacts are likely to be significant, are these likely to occur?

This same document also provides "impact significance criteria" as environmental loss and deterioration, social impacts resulting from environmental change, nonconformity with environmental standards, and the probability and acceptability of risk.

Unfortunately, the listed criteria are all challenging, if not impossible, to measure and define objectively and quantitatively, and the questions suffer from circular reasoning; that is, environmental impacts are to be considered significant if they have significant residual environmental impacts. This is another example of the struggle to quantify subjective values and inherently imprecise concepts.

6.3.1 Individual Impacts

Some environmental impact assessments stop with the identification and prediction of potential impacts and do not evaluate their significance. Of course, significance is always determined in a particular context and not in a vacuum, and it should form the basis for sound decisions. There is a major problem, however: significance is a value judgment and cannot be objectively measured. Local residents may be more interested in the jobs and income they would gain from a project, while interest groups from other areas may see major, adverse impacts from the same project. Rossouw's Table 1[26] presents nine definitions of significance from the literature. Several authors, [12, 15, 30, 33, 35], acknowledge the subjectivity of significance. For example, "determining significance is ultimately a judgment call," "significance...involves a value judgment," "the significance ...is an expression of the cost or value of an impact to society," "the evaluation of significance is subjective", and "significance is an anthropocentric concept." Other sources [5, 7, 37], governmental organizations (with one exception), imply an ability to make objective and meaningful measurements: "where anticipated future conditions ...differ from those otherwise expected," "determined within the framework of context and intensity," "significance can be considered on three levels: significant and not mitigatible; significant and mitigatible; insignificant," and "significance requires predicting change."

While the latter four definitions are accurate, they do not provide a working method of actually measuring or determining significance. What makes significance definition and measurement more difficult and complex is that the same type of project, proposed the same way (that is, with the same alternatives on similar settings) can be evaluated quite differently, depending upon the regulatory jurisdiction and its definitions. Consistency and predictability are totally lacking.

For some assessment components the evaluation of an impact is quantitative and expressed as a statutory or regulatory threshold. Air and water quality are in this category. Without knowing any particular alternative it is still possible to identify levels of air and water quality that are unacceptable because they exceed the threshold value. In the case of known toxins the evaluation has even more importance because of the potential for harm to humans, other animals or plants.

Other assessment components have impacts that are best evaluated in a larger context. Wetlands and habitats for fish and wildlife are in this category. Loss or gain of these areas has ecological meaning within the context of other similar areas, the relative sizes of the different areas, comparative habitat qualities, and distances of separation among the various locations.

The most difficult components to evaluate are those that represent an individual's values. Components in this category include æsthetics, recreational opportunities, and educational potential. Even these components, however, can be placed in a locational context. For example, a riverside resort development should be evaluated differently depending on whether the location is within a city's urban growth boundary or in a rural area. Similarly, development of a large tract of land for industrial use should be evaluated differently if it is adjacent to an established industrial area or a residential subdivision.

Separating evaluation of individual impacts from complete alternatives offers several advantages. Potential (or actual) mitigation measures can be considered on an impact-by-impact basis without the potential distraction of the cumulative effects of a complete alternative. The cumulative effects of all positive and negative impacts of each alternative are more easily evaluated and understood. The impacts can be more appropriately applied to each alternative as boundaries and a range has already been established.

6.3.2 Alternatives Evaluation

A number of formal methods for evaluating the significance of environmental impacts of alternatives has been proposed. Almost all of them include scaling, weighting, aggregating, or any combination of the three. *Scaling* is the equivalent of normalization; it puts all impacts on the same scale so that magnitudes can more easily be compared. *Weighting* "is the imposition of professional and/or societal values on a range of potential environmental impacts. This is a very contentious area, revolving around a number of issues, such as (1) Whose values should be considered? (2) How representative are they? (3) How should such values be elicited?" [26]. This part of the process cannot be too strongly stressed, for it is the basis for many legal challenges to environmental impact assessments and for dissatisfaction by one stakeholder group or another.. *Aggregation* involves combining all values to produce a composite score that can be compared among different alternatives and the baseline environmental condition. Some formal methods use an arithmetic aggregation with each value representing the same weight (or importance) as all others. Other formal methods use computerized statistical (or other techniques) to cluster similar values and separate outliers.

Reaching consensus on the value of significance of an impact is difficult, if not impossible. As noted above, *significance* is a value judgment, and while values differ one is not measurably better than another; objectively, one is not "good" and the others "bad". They are expressed in these terms because most stakeholders are frustrated by their inability to quantify the subjective. Rossouw [26] provides an overview of many formal methods and his work should be read for the details. What is pertinent here is that he discusses 24 different methods and groups them into six classes—

- Methods in which aggregation is used to facilitate comparison of project alternatives (six methods, none of which has specific provision for inclusion of public opinion).
- Methods in which there is limited consideration of impact significance (four methods and public input is part of the process in three of the four).
- Method adapted to planning (one method without specific provision for public input).
- Method with no guide on significance determination (one method with no provision for public input).
- Methods that use collective professional judgment (five methods with no provisions for public input).
- Methods that involve no consideration of impact significance (seven methods with no public input because they do not—or cannot—evaluate significance).

It is both interesting and informative that only 3 of 24 surveyed methods include specific provisions for soliciting the input of stakeholders, and these are in the category of limited consideration of impact significance. Considering the emphasis placed on societal values when considering whether the impacts of an activity are significant, the lack of public input by 87.5 percent of the two dozen widely used methods is puzzling and disturbing as well as informative. This is a serious shortcoming of the traditional approach to environmental impact assessments.

Most EIA practitioners will agree that it is highly desirable to apply each alternative to the baseline conditions in a consistent and equal way, and to do so as objectively as possible. Only in this way can the decision-makers clearly understand the choices and use them to both support the decision and to defend their choice. Many researchers and authors have proposed methods to accomplish this, but too often the practice results in extensive textual description that is difficult to comprehend and use to compare alternatives.

6.3.3 Cumulative Effects Assessment

All human activities effect the natural environment to different degrees, at different spatial scales, and at different times. The considerations of impact identification, prediction, and significance presented above are usually considered only for the project or activity being assessed. However, individual activities do not take place in isolation and the small, incremental effects of numerous projects (each insignificant by itself) can accumulate to cross the threshold into significance. Cumulative effects assessments address the larger scale concerns that all developments have on a region beyond the site of any one activity. At the highest levels of policy-making, the analyses of cumulative effects are components of strategic effects (or environment) assessments (SEA).

In the United States, cumulative effects on the natural environment are defined by the CEQ regulations as—

> ... the impact on the environment which results from the incremental impact of the action when added to other past, present and reasonably foreseeable future actions regardless of what agency (Federal or nonFederal) or person undertakes such other actions. *Title 40 CFR §1508.7.*

The spatial and temporal boundaries of the affected environment must be defined for the cumulative analyses. The components of the affected environment considered in the cumulative analyses are the same geo-physico-chemical, biological, and socioeconomic components that are considered for the specific project or activity in the assessment. The spatial limits of the cumulative analyses may be broader than the project site because the former area considers all activities that affect those environmental components. As examples, if storm (or process) waters are treated to meet discharge criteria before the waters leave the project site, then there is no contribution to cumulative effects for water quality. But migratory animals, such as mule deer, may be affected by the project within a 10-kilometer radius of the project site; however, during migration, the mule deer may be affected by other activities in the areas through which they migrate. Therefore, those activities must be considered in the cumulative analyses. While other activities must be considered, the geographic boundaries must not be extended to the point that the analyses become unwieldy and not useful for practical decision-making.

The temporal boundaries of the cumulative analyses often extend beyond the period of time considered in the analysis of the specific project alone. This is because the cumulative analyses must take into account activities that occurred prior to the start of the current activity as well as after the proposed project is completed. The time scale must be for the same resources considered in the specific project being assessed. A major shortcoming with the temporal scale of cumulative effects analyses is seen when mathematical models of processes are run on a computer. For example, just because a hydrogeological (i.e., groundwater) model can be run for a 300-year period does not mean that its predictions of water quality in an adjacent lake are valid that far into the future. Climate and human population changes over that same three centuries will affect water quality, too, but are not included in the model. *It is extremely important to limit the temporal boundaries to times that are both reasonable and realistic.* To understand the limitations of forecasting environmental change better look carefully at weather predictions for only very short periods of time (one week, for example) and compare that to the actual weather. There is a high degree of variability in prediction accuracy because of both the inherent variability of natural systems (including weather) and the ability of predictive models to include all the pertinent variables and to assign the proper values to constants.

Germane to both spatial and temporal boundaries for the cumulative effects analyses, reasonable assumptions must be made about additional activities even if there is uncertainty regarding them and their potential effects. *The specificity of the analyses should be consistent with the nature, scales, and degrees of uncertainty regarding the particular activity being evaluated.*

While cumulative analyses may help to place a specific project within the broader landscape, such broader views and also the basis for Strategic Environmental Assessments (SEA). An SEA involves reviewing policy, plan and program proposals to incorporate environmental considerations into the development of public policies. Reiterating how subjective are the principal issues of impact assessments ("significant," "acceptable," "reasonable"), it is a logical conclusion that policies based on these imprecise concepts are also likely to be contentious.

7

Writing the Impact Statement

There may be many problems with how environmental impact assessment reports are written. This is not the place to warn against sloppy, inconsistent writing; poor grammar; and other editorial ills (for that, see [4]). This is the place to note easily corrected major deficiencies that can worm their way into reports.

With small differences among jurisdictions and level of the assessment, there are certain chapters (or sections) common to most environmental impact assessment reports. The key sections are Executive Summary, Statement of Purpose and Need, Description of the Proposed Project, Existing Environment, Description and Comparison of Alternatives, Description of the Affected Environment, and Environmental Consequences. As noted in [14], the report is an "aid in communicating information to both technical and nontechnical audiences."

The baseline conditions (in the Existing Environment chapter) should be described not only in text and well-planned graphics (maps, plots and other figures) but should answer the question, "So what?" That is, the report needs an objective assessment of the current conditions, not the implicit assumption that those conditions are already in a highly valued state or one that is desirable by all stakeholders. Virtually all land has been altered directly or indirectly by natural and human forces. Being candid with the baseline condition builds credibility in the rest of the report. For example, if wetland areas have been planted in pasture and used to graze cattle, the condition of the ground and vegetation needs to be objectively expressed. In this case, the area may well be jurisdictional wetlands, perhaps prior converted wetlands, but a technically sound evaluation of the wetland functions and values may show them to be degraded. In such a case, the description of baseline conditions should indicate the degradation and not just report the number and size of the wetlands.

The proposed project site also needs to be placed into a regional (or, at least, broader than local) environmental context when describing the baseline conditions. This means that the environmental conditions for the spatial and temporal extent included in the cumulative effects assessment need to be described. The rationale for the establishment of the spatial and temporal boundaries should also be provided in this chapter.

Alternatives need to be described and analyzed equally. This means that the description of each alternative should cover the same topics to the same depth. The "No Action" and Preferred Alternative should not stand out by being treated superficially, or by being the subject of a sales presentation. The report is the basis for making a decision on the project and is the main forum for presenting the project to the public and the decision-makers. When alternatives are evaluated for their potential effects on the baseline environment, the topics covered and the analyses used must be the same for each one. It is not adequate to present the "No Action" alternative as not changing the existing environment, for those conditions will definitely change over time regardless of whether the project is approved and undertaken.

The Preferred Alternative, too, should not be treated differently from the others. There may well be large negative impacts associated with the project, but they may be mitigated, or at the end of the project the net benefits may well outweigh the negative impacts. The only way to demonstrate that the project is worth approving is to be completely objective and equal in attention to each alternative.

The most important part of the report is a complete description of how impacts were identified, predicted, and evaluated for significance. Because not all input data can be measured—or measured adequately or for long enough to satisfy academic scientists—the report needs to explain where qualitative data were used and how these data were evaluated, particularly in comparison with quantitative data.

Part II

The Modern Approach

8

Moving to the New Paradigm

In the first part of this book environmental impact assessments were described procedurally; that is, in sequence from initial public scoping through data collection and analyses, alternative impacts, and writing the report. There are other ways of organizing and describing the environmental impact assessment process. One alternative is to look at core issues found among multiple components regardless of where in the assessment process they appear.

Core issues include—

- Ranking preferences.
- Quantifying significance, acceptability, sustainability and other societal values and beliefs.
- Describing environmental conditions by combining quantitative and qualitative environmental data to summarize and characterize these conditions.
- Impact inference, evaluation and assessment.
- Making multi-objective, multi-criteria decisions under conditions of uncertainty.

8.1 Tools for the New Paradigm

The tools now available to the EIA practitioner and decision-maker are many. All can address core issues and provide the basis for making a decision. Some of these tools are methods that use cardinal values (numbers) and some are methods that use ordinal values (words).

Some methods are computationally intensive and require a computer; other methods allow calculations using a spreadsheet or by hand while providing approximations adequate for many situations. All of these methods transform subjectivity into objectivity by an open, explainable process. The manipulations are mathematically sound and based on well-established theory and research.

8.2 Progress Toward Adoption

Environmental impact assessment practitioners and researchers have recognized the value of the modern approach. Various components and approaches have been used in a range of project types. Five such projects are summarized to illustrate the broad recognition that a new paradigm is necessary.

8.2.1 Mercury Bioaccumulation Risks

Viega and Meech [38] applied fuzzy logic to assess the environmental risks of mercury bioaccumulation associated with gold mining in operations in Brazil's Amazon region. As they point out, the fuzzy system model provides the same practical risk level results as a complex mathematical model but without the high costs, large amounts of data, and sophisticated technical skills to relate factors and bioaccumulation quantitatively. The authors developed a well-thought-out fuzzy expert system they call "HgEx" that "accommodates imprecise data input for variables, such as background level as well as how measurements are transformed into linguistic expressions with respective degrees of belief to be handle [sic] in a heuristic model (neural equations—Weighted Inference Method)."

Their approach uses a combination of a neural network[1] and a fuzzy expert system (called a *neuro-fuzzy* model) to address the problem. This approach uses a simple neural network to propagate weighted data evidence to a conclusion that forms the input to the approximate

---

[1] A neural network is a computer model that mimics the human brain's ability to identify patterns. While the outcome of the neural network can be proven correct it is not possible to determine exactly how the model derived that outcome.

reasoning (IF-THEN rules) inference engine. This heuristic approach builds the model rules from data collected for the assessment rather than from pre-conceived ideas of what will be present at any given site. Part of the complexities they had to overcome are the differences in mercury affinity for different sediment types and how this affects bioaccumulation potential.

They classified the large diversity of sediment types into three groups, each representative of low, medium, and high mercury. They then created rules[2] such as:

```
IF sediment is classified as Type_1
   THEN Hg_level is low
      AND certainty_factor is very high.
```

and

```
IF sediment is classified as Type_1,
   AND hydrous ferric oxides are present,
   AND sulphide mineral is present,
   AND organic matter is present
      THEN Hg_level is low
      AND certainty_factor is zero.
```

The first rule states that one can have very high confidence that mercury levels are low in Type 1 sediments (gravels, white or light gray clay or sand, limestones and sandstones). However, if there are other factors present that are known to facilitate mercury bioaccumulation, the confidence is zero that the sediment type will contain low levels of the metal.

8.2.2 Checklist Enhancement

Anile and Gallo [1] created a computational intelligence extension to the Battelle method checklist (Section 4.1 on page 28) that uses fuzzy arithmetic to create a rating matrix. The rating matrix is designed to measure the "degree of the impact of the n-th effect caused by the m-th action." Their paper describes how to assign a numeric entry to each

---

[2] These are explained in detail in the following chapter.

matrix cell, how to aggregate the cell values into an overall impact estimate and how to decide among various project alternatives based on the completed rating matrices.

The rating matrices are created using a two-step process. First, a group of experts is asked to rate each element's degree of impact using a set of prescribed terms for impact sign (positive, negative, neutral), importance (great, middle, small), probability (certain, probable, improbable), and duration (temporary, permanent). In the second step the analyst assigns a number in the range [0,1] according to a defined scale. The results are represented as triangular fuzzy sets and aggregated by either an averaging or geometric mean method. The authors believe that this method provides a sensitivity analysis of the environmental assessment and the selection of the most appropriate alternative based on the least sensitivity of the environment to that alternative project approach.

8.2.3 Fuzzy Logic GIS

A hybrid approach incorporates spatial analyses (using a geographic information system [GIS]) with fuzzy logic. Such a hybrid system is particularly useful for large-scale (i.e., landscape) characterizations and assessments ([25]). These fuzzy-GIS systems are excellent for planning and management as data are accumulated over time. However, the input data requirements and assumptions about spatial changes associated with different alternatives for a project-specific environmental impact assessment may limit their use in the latter applications.

8.2.4 Fuzzy Decision Analysis

An interesting combination of techniques was put together by [36] to create a fuzzy decision analysis method for integrating ecological indicators at the drainage basin scale for a large portion of the mid-Atlantic region of the United States. The problems they had to solve included integrating information from many different sources into an overall ranking of ecological risk relative to each basin. Using 26 indicators of regional ecological conditions, they converted measured values to fuzzy grades of membership. These fuzzy numbers were converted to distance measures calculated by fuzzy arithmetic. The distances were ranked by magnitude, clustered by principal components analysis (PCA), and input into an Analytical Hierarchy Process that produced composite scores for each of the 123 drainage basins.

8.2.5 Fuzzy EIA

An effort to apply fuzzy logic to significance determinations in environmental impact assessments used some aspects of fuzzy sets and fuzzy logic but attempted the calculations manually [19]. This publication compares an environmental impact assessment conducted qualitatively for a project in Canada with one conducted semiquantitatively in Mexico. Both assessments relied primarily on professional judgment of the author or other experts. Unfortunately, however, instead of using measured values of magnitude (intensity), spatial extent, duration and other impact criteria, they assigned an ordinal number (0–4) to the categories of none, very low, low, high and very high. It was these ordinal numbers that they then used as input values to triangular fuzzy sets. As will be shown in the following chapter, this is not how to capture the uncertainty of the measurements as grades of membership in fuzzy sets. When the modern approach is misapplied through lack of sufficient understanding, the underlying ordinal (descriptive) input data are hidden behind a veneer of quantitative manipulation and appear more rigorous than is justified. As will be shown below, linguistic variables such as magnitude of intensity can be directly converted to a grade of membership in a fuzzy set. Linguistic variables of spatial extent and duration are expressions of areal and time measurement and are very easily converted into a fuzzy grade of membership.

These examples illustrate different paths to resolving some of the subjectivity, complexity and variability inherent in ecosystem description and risk analyses. None of the above, however, describes a complete environmental impact assessment as it is usually required by statute and regulation.

Introduction to Fuzzy Sets and Logic

9.1 Measurement and Language

Every language has words that describe measurements regardless of the system used (e.g., Metric, English, Apothecary). People communicate measurement information using objective, or "crisp," language; for example, the distance between city A and city B is 592 kilometers; this wetland has an area of 4.3 acres; that medicinal pill contains 45 grains of active ingredient. Depending on the measuring device, resolution differences are identified at greater or lesser scales.

Languages have other words for measurements, words that do not crisply define magnitude. These words capture the concept of relative, rather than exact, amount. For example, city A is *far* from city B, the wetland is *small,* the pill contains a *large* amount of medicine. These subjective, *linguistic variables* are imprecise, vague, and *fuzzy.* Everyone has a concept of *far, large, expensive* and *heavy* but the magnitude each person assigns to the terms differ. And, what is meant by *large,* for example, varies with the context. A large meal is measured on a different absolute scale than is a large house. The result of all this fuzziness leads to confusion and misunderstanding unless the fuzzy linguistic variables are quantified.

Fuzzy linguistic variables also include those concepts that do not have an underlying measurement. Everyone understands what is meant by a *beautiful* sunset or an *acceptable* price. But, we cannot describe what we mean very easily or consistently. Both types of fuzzy linguistic variables (those having an underlying measurement and those that are purely abstract) are found in environmental impact assessments.

9.2 From Subjectivity to Objectivity

The subjective nature of *significance*, the inability to collect very large data sets on baseline conditions, and the uncertainties about future conditions under a set of project alternatives are all reasons why traditional approaches to environmental impact assessment should be replaced with tools that use advances in mathematics, the increased power of small computer systems, and mature aspects of artificial intelligence.

Among the techniques within the broad category of computational intelligence are fuzzy logic, approximate reasoning (IF-THEN models), evolutionary/genetic algorithms (the terms are used interchangeably), artificial neural networks, Bayesian-based reasoning (using expectations based on past experience), belief networks and Dempster–Shafer theory of evidence-based reasoning.

Approximate reasoning is a mathematical subject "with emphasis on the design and implementation of intelligent systems for scientific and engineering applications. Approximate reasoning is computational modeling of any part of the process used by humans to reason about natural phenomena" (from the mission statement of the *International Journal of Approximate Reasoning*). In other words, it is ideally suited to be a basis for a modern approach to environmental impact assessment. Put another way, approximate reasoning is concerned with formal models of reasoning under uncertainty.

In order to understand why fuzzy sets, fuzzy logic, and approximate reasoning can overcome the shortcomings documented in the first part of this book, and to understand how to use fuzzy system models most effectively in conducting an environmental impact assessment, some background is necessary. The mathematics will be minimal and are there for those who want this level of understanding; most of the explanation is effectively communicated by text and drawings. This chapter is not a comprehensive explanation of fuzzy sets, fuzzy logic, or fuzzy system models. For more depth and completeness see [3, 9, 20, 45]. The contents of this chapter provide basic understanding of principles, techniques and language that underlie their use in conducting environmental impact assessments with enhanced objectivity, mathematical rigor, and soundness.

9.3 Linguistic Variables

When uncertainty or imprecision is with the word used rather than with the event, that uncertainty or imprecision can be addressed by fuzzy logic. Examples of lexical imprecision are the concepts of *steep* slope or *significant* impact. This imprecision reflects the ambiguity of human thinking when perceptions and interpretations are expressed. "The project area is a healthy ecosystem" and "the noise levels are loud" are examples of statements that are inherently fuzzy. *Linguistic variables* in the previous sentence are "Ecosystem_quality" and "Noise_-level", respectively. These variables contain descriptive *fuzzy terms*, such as "healthy" and "loud," that represent a range within the variable. More familiar are the fuzzy terms *young, middle-aged*, and *old* as divisions of the linguistic variable "Age."

Linguistic variables are ideally suited to express the concepts found in environmental impact assessments. For example, the linguistic variable "Significance" can be defined by the term set "Highly_insignificant," "Insignificant," "Moderately_significant," "Significant," "Very_-significant"; this collection of fuzzy sets is called the *term set* for Significance. The number of fuzzy sets in the term set is not fixed but generally varies from three to seven sets that overlap (usually by 50 percent) in the values they include. The scale used to measure linguistic variables is determined by need. The range, from lowest value to highest value of all term sets (along the x-axis), is called the *Universe of Discourse* (UoD); that is, it is the numeric range of a variable that spans the term set. The range of each term in the set that composes the UoD is determined by the lead agency, the environmental impact assessment preparer, or by consensus to reflect local values and beliefs. The x-axis value range of a fuzzy set is called the *support set* for that term. Interpreting the meaning of a dynamic fuzzy set is enhanced by recognizing the difference between the support set and the Universe of Discourse.

Numerous linguistic variables appear in environmental impact assessments, including concepts of size, distance, number, pristineness, significance, sustainability, and acceptability. While quantitative measures of the first three are possible, the meaning of that measurement in the context of the environmental impact assessment is not at all objective. Immeasurable concepts such as pristineness, significance, sustainability, and acceptability reflect the value of a natural system component by different stakeholders and interest groups and are not directly measurable. The difficulties of quantifying immeasurable societal values extends beyond environmental impact assessments to the related processes of natural resource damage assessments (NRDA) and environmental risk management.

9.4 Ranking of Preferences

There are several broad categories of ranking applicable to fuzzy numbers, fuzzy sets, and ordered-weighted averaging aggregators (OWAs). In environmental impact assessments ranking plays a role at several stages; as examples, during the scoping and decision-making processes.

In the scoping process potential project alternatives and baseline environmental components are identified, screened for relevance and ranked by subjective value judgments of different stakeholder groups. The foundation for this ranking is pairwise comparisons of alternatives or environmental components. Deciding which of two choices is more desirable, and by how much is not always easy. For this reason, one option is that the two are essentially equal in importance to the decision-maker. Focusing on only two choices at a time is much more manageable than trying to rank a large number of choices simultaneously. There are too many competing values to produce meaningful—or useful—results. However, selecting one of two choices focuses on their specific attributes and can be done by everyone who wants to participate in the process.

When this pair-wise comparison of alternatives or components is performed by every stakeholder interest, all results are equally treated. This means that the values of each group that compared a to b is averaged (or aggregated by another method) and placed in the appropriate cell of a square matrix. It is reasonable to expect that within a group of stakeholders with diverse interests, all compared pairs will have a range of values by which one of the pair is more important than is the other. Among these stakeholders, some will consider *a* the more important, and the rest will consider *b* more important. The key aspects of this process are that everyone who wants to express a preference can do so, the strength of each expressed preference is from the same numeric range, and the conversion of the matrix into an ordered ranking is mathematically robust and completely objective. Given the above, soliciting values and beliefs from the broadest spectrum of stakeholders has major benefits to the project developer. A common complaint, that the views of some stakeholder were not allowed to be expressed or included in considerations by the lead agency, is removed by this approach.

When alternative actions are evaluated, the focus of the decision-maker is on ranking a set of alternatives relative to a limited number of decision criteria. To do this, performance values of the alternatives for each criterion must be known, i.e., how each criterion affects each alternative. When there are multiple criteria (or constraints on achieving an alternative objective), two situations are possible: all criteria have equal importance toward the objective or they have different importances. While the former case is easier to compute, the latter situation is much more realistic. It is often necessary to guide the decision-makers to consider how these criteria differ; it is not necessarily immediately obvious to them. If the criteria weights may change, the number of possible rankings is dependent on the numbers of both alternatives and criteria, but this number is less than the factorial of the number of alternatives [?]. This situation occurs both when different decision-makers in a group have different weights for the criteria and when it is desired to provide a sensitivity analysis of the ranking process.

A formal method of ranking project alternatives (objectives) when each alternative is constrained to different degrees by criteria has another benefit in the assessment of environmental impacts: existing conditions are evaluated relative to alternative future conditions. The relative rank of the existing conditions to the predicted future environment under the "No Action" alternative provides important insight to the decision-maker and strengthens support for the chosen alternative.

A technically-sound ranking method facilitates both the making of a decision and supporting the choice. When subjectivity is removed and values quantified, decisions tend to be less arbitrary or capricious. When a project is particularly contentious, the greater the objectivity and ability to audit how a decision was reached, the higher the likelihood of acceptance by all parties. Of equal importance, the process is quicker and less costly to complete. A final ranking of alternatives can also be used when economic, social, or environmental conditions change prior to project initiation and a different alternative needs to be adopted. Since the work has already been completed selection of a different alternative is expedited and justified.

9.5 Fuzzy Sets (Type-1)

Fuzzy sets are best explained, and understood, when compared with the more commonly understood concept of a crisp set. The crisp set concept was developed by the mathematician Georg Cantor. Such Cantorian sets are evaluated and combined using truth tables based on the symbolic logic rules devised by George Boole and called Boolean logic.

9.5.1 Crisp Sets

Most of us are educated to think about numbers as things we can count and measurements we can take. The number of plant species in a wetland or the amount of dissolved oxygen in a stream are examples of directly counted or measured values. These are called *crisp* values because there is no ambiguity—other than measurement imprecision or instrument error. The number of plant species in the wetland is an integer; there will not be a fractional number of species. If there is uncertainty of the identification to the species level it is a measurement uncertainty, an imprecision in our ability accurately to determine the plant morphometry or characteristics. Similarly, the accuracy and precision of the dissolved oxygen meter, the time when the measurement was made and variations with depth and other spatial parameters result in an estimate of how "good" is the reported value. None of these uncertainties or imprecisions are inherent in the variable being counted or measured. Rather, they are reflections of the how we count or measure them. The attribute of crispness also applies to sets as well as to individual measurements; for example, the set of dead trees on the project site is a number determined from counting all the dead trees.

Sets are collections with an attribute in common. As examples, the set of coniferous trees or the set of identified wetlands on a project site. Crisp sets are collections of distinct objects. In every situation, a new object (set element) can be tested to learn if it is a member of a specified set. Mathematically, this testing is done by the characteristic function of the set. The characteristic function $\mu_A(x)$ of a crisp set A, in U, takes its values in the range (0,1) and is defined as

$$\chi_A(x) = \begin{cases} 1, & \text{if and only if } x \in A \\ 0, & \text{if and only if } x \notin A \end{cases} \tag{9.1}$$

The numbers 1 and 0 are interpreted as A belongs in the set x or A does not belong to set x. These two numbers are often written as {0,1}. Using this brace notation, (9.1) can be expressed as

$$A = \{(x, \chi_A(x))\} \tag{9.2}$$

where $(x, \chi_A(x))$ is called a *singleton*. Singletons also have a place in fuzzy sets and this role will be explained later.

A graphic example of a crisp set is a steep slope, when *steep* has been defined as a grade of 20 percent or more (Figure 9.1 on the following page). For this set let U be the real line $\Re^1$, and let crisp set A represent "steep slopes whose grades are real numbers greater than 20 percent". Then $A = \{(x, \chi_A(x)) | x \in U,$ where the characteristic function is

$$\chi_A = \begin{cases} 1, & x \geq 20 \\ 0, & x < 20 \end{cases} \tag{9.3}$$

Because the only two possible values are 1 and 0, there is an excluded middle (all values between these two extremes). Crisp set thinking leads to the well-known conundrum of whether a glass in which 50 percent of the available volume is filled with water is half-full or half-empty. More importantly, the problem from a regulatory perspective is such a crisp threshold value can be meaningful. If development on "steep slopes" (i.e., slopes of 20 percent or greater) is prohibited, is it permissible to develop on a slope of 19.5 percent?

Within the traditional environmental impact assessment process there are two reasons for crisp thresholds: they are unambiguous and there have not been alternatives until now. Crisp thresholds and crisp sets were adopted to address concepts of risk associated with industrial and developmental projects. Risk may be expressed as "significant

---

[1] That is, the set of all real numbers.

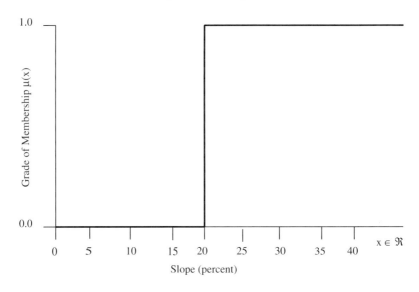

Fig. 9.1. The crisp set of steep slopes. The threshold is at 20 percent. All grades greater than 20 percent are members of the set; all grades less than 20 percent are not members. Exactly 20 percent is an ambiguous discontinuity.

impact" or "undue degradation" or the perception of societal or political "harm." The outcome is the statutory and regulatory establishment of sets to be used for project evaluation. Examples of such risk-based sets include standards for water quality, slope stability, and minimum sustaining population sizes.

Sets are extensively used within environmental impact assessments. The scoping process produces the set of environmental components that will be considered in the assessment. The baseline conditions comprise the set of environmental components that exist on the project site. The alternatives are evaluated on their impact to the component sets individually and collectively. These sets are all crisp; that is, their membership is either 1 (a member of the set) or 0, not a member of the set. Either a slope has a grade less than 20 percent (and is, therefore, not steep) or it has a grade equal to or greater than 20 percent (and is, therefore, steep). Sets change from crisp to vague when the meaning of the measured values is considered. Examples of this transition occur when attention is changed from the measured slope of the ground to whether that slope can be characterized as *shallow, moderate,* or *steep.*

9.5.2 Fuzzy Sets

In 1965, Lofti Zadeh [42] at the University of California, Berkeley, created the mathematical theory and tools to quantify linguistic concepts, words that have meaning but are inherently imprecise, vague or fuzzy. Each fuzzy set is defined by a *membership function* that is used to calculate the *grade of membership,* and these sets are rigorously manipulated using the tools of *fuzzy logic.* Rather than either being a member of a crisp set (e.g., *tree*) or not being a member of that set, Zadeh showed that it is possible to be a member of a set, for example, *short* (expressed as $\mu_{short}(x)$) as well as a member of the set *tall* (expressed as $\mu_{tall}(x)$). While fuzzy sets can be defined by different shapes, a generic fuzzy set–with all parts labeled—is shown in Figure 9.2.

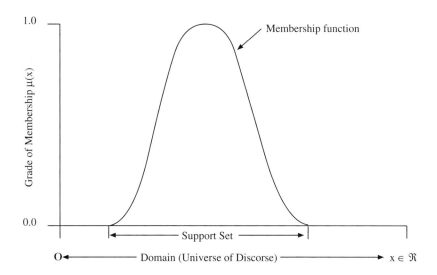

Fig. 9.2. A generic fuzzy set showing all the parts discussed in the text.

The meaning of the grade of membership, μ, can be difficult to understand. In most applications the grade of membership ranges from

[0.0] to [1.0]. The value of μ is not a measure of probability, but in some models it may represent a possibility. The meaning of μ in the context of environmental impact assessments is that of truth; truth of the assertion that a measured value is a member of that particular fuzzy set to that degree (or grade). Thinking of grade of membership as a truth value is easier in the context of fuzzy sets being operated on under rules of a formal logic. Fuzzy logic is a formal logic that is based on truth tables, just as Boolean logic is applied to assess the truth of a value being a member of a Cantorian (crisp) set.

Fuzzy sets quantify vagueness in membership value because they have a gradual transition between nonmembership and membership. The relationship between crisp and fuzzy sets is that the former are a special case of the latter. A fuzzy set, $\overrightarrow{A}$, in the range of values, U, can be defined as a set of ordered pairs,

$$\overrightarrow{A} = (x, \mu_{\overrightarrow{A}}(x))|x \in \overrightarrow{A} \qquad (9.4)$$

where $\mu_{\overrightarrow{A}}(x)$ is the *grade* (or degree) *of membership* of x in $\overrightarrow{A}$; that is, the amount by which x belongs in the set $\overrightarrow{A}$, or the degree of truth that x is a member of set $\overrightarrow{A}$. The shape of the membership function reflects the semantic meaning of the term and the range of measured values across which the membership function has a grade of membership greater than zero is not the same as the total range of values that can be measured. The meaning of membership function shape and the other labeled features is explained in Section 9.5.4 on page 77.

Two types of linguistic variables are needed in environmental impact assessments. One type represents the grade of membership of a number in a fuzzy set; e.g., a measured value such slope grade (Figure 9.3 on the next page). In addition to slope grade, stream length, wetland area, population size, traffic volume, noise level and number of jobs are all measurable values that produce a number whose meaning is best expressed as a grade of membership in a fuzzy set. This type of fuzzy set quantifies the subjective terms applied to these measures (e.g., *large, many, far, heavy*) and is referred to as a Type-1 fuzzy set.

The second type of fuzzy set is, arguably, even more useful because it permits the quantification of values, beliefs and inherently imprecise or uncertain terms such as *significant, acceptable, unpleasant, sustainable,* and *æsthetic*. These are pure linguistic variables, but they can be quantified when a suitable scale and membership function shape is used. An

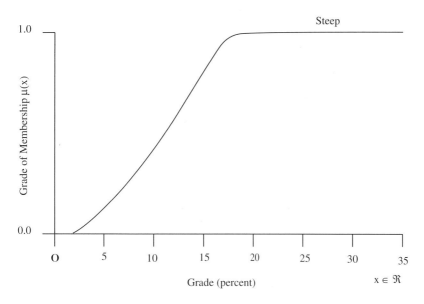

Fig. 9.3. The fuzzy set of steep slopes. Each measured grade > 2 percent is a member of this set to some degree.

example of a concept-based fuzzy set is shown in Figure 9.4 on the following page and is called a Type-2 fuzzy set. It is important to understand this example as it relates to design decisions of such fuzzy term sets, particularly when it is necessary (or desirable) to work strictly with Type-1 fuzzy sets. In this case, the lead agency, environmental impact assessment practitioner or knowledge engineer worked with a representative sample of stakeholders and technical participants in the assessment process. During discussion of the meaning of the concept of *significance*, everyone agreed that a cumulative score greater than 25 (on an arbitrary scale) was unquestionably *significant*. Below that value a reasonable first approximation is that the change in grade of membership is linear and rapid. It is equally valid to use an S-shape curve to represent a concept like *significant* if the prevalent view is that the grade of membership increases comparatively rapidly between [0.0] and [0.5], then more slowly as it asymptotically approaches [1.0]. There are no firm rules except that the shape of the membership curve reflects the semantics of the concept. The greater the agreement among

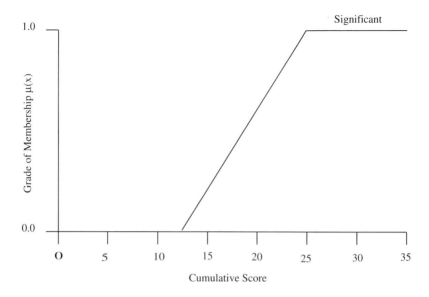

Fig. 9.4. The fuzzy set, *significant*, demonstrating that the grade of membership varies until there is consensus that all higher values are completely within the set.

stakeholders that the representation is fair, the greater the power of the output to support decisions. The better approach, using Type-2 fuzzy sets, is explained in Section 9.6 on page 85.

Fuzzy sets have two important properties. The first property is that an element has value both by its own grade of membership and by that grade's position relative to other elements in the set. In the set of steep slopes shown above (Figure 9.3 on the preceding page) a 15 percent membership [0.87] has meaning by itself, but even more meaning relative to the 10 percent slope's grade of membership [0.41]. The rate of change in degree of membership associated with a change in slope grade describes the truth trajectory of the concept. The second property relates to the errors of including, or excluding, a particular element in a set. With crisp sets, as the value of the element approaches the threshold (from either side), the uncertainty of whether that element is in or out of the set increases sharply. With fuzzy sets, this inclusion/exclusion error is spread uniformly across the entire support set; there is no discontinuous value where it is impossible to determine set membership.

So far, a Type-1 fuzzy set has been a single term that is descriptive of a measured number or, under certain circumstances, a word. However, not all slopes will fall within the support set of *steep*; other slopes might be *flat*, *slight*, *moderate*, or *extreme*. Each of these terms is a fuzzy set in the collective term set for the linguistic variable "Slope_steepness" (Figure 9.5). What sets a term set of Slope_steepness apart from the

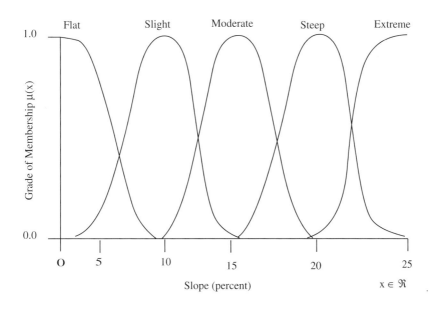

Fig. 9.5. A fuzzy term set for slope steepness.

crisp set is the ability of a slope to have a grade that falls in two categories. For example, a slope measuring 13 percent has a degree of membership of [0.65] in the fuzzy set *slight* and a grade of membership of [0.35] in the fuzzy set *moderate*. This is much more realistic than the crisp threshold of a defined percentage grade.

9.5.3 Fuzzy Numbers

There are situations in which a crisp number needs to be used in a fuzzy system model. Since there is no mechanism for mixed-mode use, the crisp value needs to be converted to a fuzzy set. This fuzzy set is called a *singleton*. This singleton is equivalent to the crisp set single-ton described above. The fuzzy set singleton is created as a set where only the number itself has a grade of membership of [1.0] and all values on either side of that number have a grade of membership of [0.0] (Figure 9.6). Now this fuzzy set can be combined with other fuzzy sets

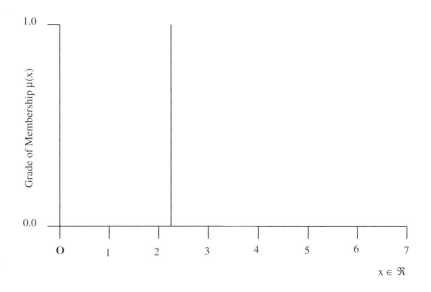

Fig. 9.6. An example of a singleton fuzzy set for the number 2.25.

with fuzzy operators in a rule-based system.

A fuzzy number by itself is frequently too constraining. Consider the situation of sloping ground. With the exception of sheer rock faces and a few other landscape features, hill slopes have varying grades from bottom to top. Even the most accurate survey will produce an average value for the overall slope or any portion thereof. .In the application of fuzzy numbers to the natural world we need qualifiers. Qualifiers are terms applied to a number, such as "around," "about," and "near." It is usually more appropriate to convert the fuzzy single-

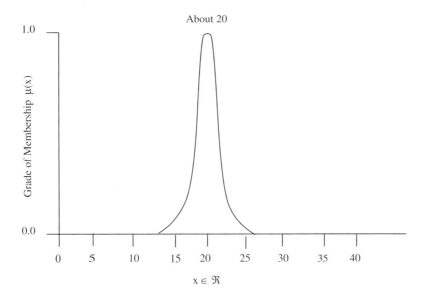

Fig. 9.7. One way of expressing the concept of "about 20." The shape and width of the membership function will vary with the meaning in different situations.

ton into a fuzzy set by applying a qualifier. Figure 9.7 illustrates how the slope grade *about 20 percent* could be represented. With numeric fuzzy sets and various qualifiers every type of number and measurable value can be represented as a grade of membership and used in a fuzzy system model.

9.5.4 Membership Function Shape

The shape of the curve used to describe the fuzzy set, and from which the grade of membership is calculated, needs to reflect the semantic meaning of the term. There are three basic shapes for fuzzy sets: triangular (Figure 9.8 on page 79), trapezoidal (Figure 9.9 on page 80) and bell-shaped (Figure 9.10 on page 81). The semantic meaning of the concept described by the fuzzy set is how membership in the set changes as the measured value (the independent variable) increases. In all three cases, the grade of membership increases along the y-axis as the measured value moves from left to right along the x-axis. Selecting the appropriate shape of the membership function is crucial to results that most accurately reflect the relationship between the data and the underlying concept. Therefore, while experimental evidence suggests that shape can often be inconsequential, the membership function must be deliberately and thoughtfully designed.

A linguistic variable is represented by a term set rather than a single fuzzy set. To capture the semantic meaning of the variable as accurately as possible consideration is given to the shape of each fuzzy set (the ones on either end of the Universe of Discourse may be S- or Z-shaped or a shouldered trapezoid), the number of fuzzy sets within the variable, the range of each one (the support set) and the degree of overlap (50 percent is a good starting point).

Control and engineering applications of fuzzy logic (e.g., elevators, anti-lock brake systems, image stabilization for video cameras) generally use triangular or trapezoidal membership function shapes. There are no subtleties to the underlying semantics: the grade of membership increases linearly until it equals [1.0], then it decreases linearly back to [0.0]. The triangular fuzzy set divides the Universe of Discourse into linguistic pieces meaningful to the control of a machine or a process.

Within expert systems such as that used to quantify environmental impact assessments, triangular and trapezoidal fuzzy sets are not extensively used for measured linguistic variables. A triangular fuzzy set might be used to create a fuzzy number from a measured, crisp value, but otherwise it does not represent the underlying semantics of the variable. For example, the fuzzy set for "about 20 percent slope" may range from 15 to 25 on the x-axis. The grade of membership may decrease quickly on either side of 20. In this situation, a triangle may be an appropriate shape. On the other hand, the meaning of "about 80 percent slope" suggests that grades close to 80 percent have almost the same degree of membership. In this case, a bell-curve where the grade of membership decreases gradually to [0.5], then more rapidly to [0.0] will more closely represent the slope of the land. Depending on what the number represents, the semantic meaning can be best represented by either a triangular or bell-shaped fuzzy set. The factors involved in the decision of what shape to use for a fuzzified number include the range of measured values on either side of the crisp number that is included in the set and the relative importance of distance away from the

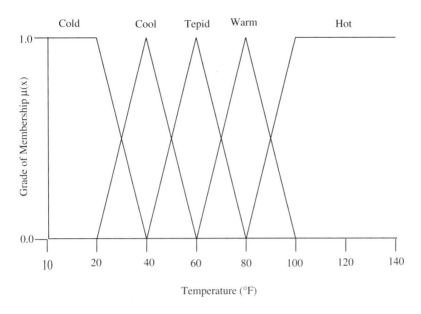

Fig. 9.8. A triangular fuzzy term set.

crisp number. With the examples given above, the apparent differences between 18–22 percent slope may be perceived as greater than the difference between 78–82 percent slope, although each is a 4 percent range around the central value.

The trapezoidal fuzzy set represents a value that increases linearly from [0.0] grade of membership to [1.0] over a limited range of values. The membership remains at the grade of [1.0] for a range of support set values before decreasing linearly back to [0.0]. This form might be used to represent concepts such as "acceptable," "significant," "sustainable," or other variables that cannot be measured or assigned a narrow range of values where the grade of membership is [1.0]. Even partitioned into a term set of three fuzzy sets (e.g., *Insignificant, Significant, Highly_significant*), a grade of membership of [1.0] will span a range of values within each set. Most people would be comfortable with this expression of these ideas because they cannot be more precisely defined.

The sigmoid and bell-shaped membership curves are extensively used in environmental impact assessments to represent the meaning

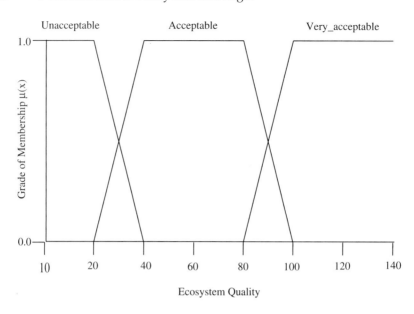

Fig. 9.9. A trapezoidal fuzzy term set

of measured components in the existing environment and changes predicted under different alternatives. Because there is never enough time or money to collect sufficient data to completely characterize the ecosystems in and around a project area, generalization is both appropriate and more accurate than using raw numbers. This is particularly true when combining subjective estimates of quantity and quality with measured values. All surveys and measurements of the existing environment represent a snapshot in time. For some important components, for example, noise and odor, the numbers produced by measurement instruments do not represent the effects on people and other animals. Therefore, fuzzifying measured numbers and qualitative assessments greatly enhances their meaning and makes it much easier for decision-makers to interpret those numbers. All fuzzy term sets should represent local values and agreement among stakeholders and other interested parties. A change in measured value may have either a small effect on membership grade or a large effect, depending on the magnitude of the value, the number of fuzzy sets and their shape. These

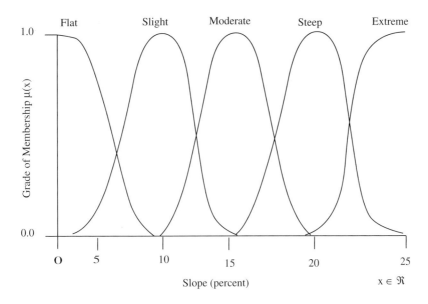

Fig. 9.10. A bell (sigmoid, S)-shape fuzzy term set.

are adjustable for each assessment so they best capture and reflect the values of everyone involved in the process.

There are no fixed shapes for fuzzy sets. Each set maps the grade of membership to the concept being modeled. As the domain is traversed from left to right, the grade of membership increases to the maximum of [1.0], then declines to [0.0]. It is up to the practitioner to develop the most appropriate shapes. The only constraint is that each fuzzy set in the term set of a linguistic variable must be *normal*; that is, it must include [0.0] and [1.0] as grades of membership. A sigmoid fuzzy set (S- or Z-shaped) has only one end at each of these values, a bell curve will have both ends at [0.0] and the middle at [1.0]. The advantage of a bell-shaped membership curve over a triangular shape is that the former has an inflection point on each side. Below the inflection point, the grade of membership decreases quickly with a change in the support set (until the toes of the curve) while above the inflection point the grade of membership approaches (and departs from) the maximum value slowly and asymptotically. These two responses fit a common perception of how environmental conditions change with time, place or quality assignments.

9.5.5 Alpha Cuts

For many fuzzy sets there is a minimum grade of membership below which there is no meaning (Figure 9.11). This grade of member-

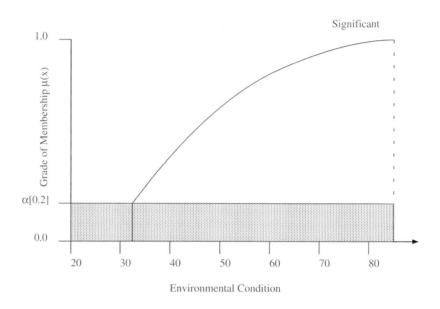

Fig. 9.11. Alpha-cut threshold for fuzzy set *Significant*.

ship is called the alpha-level or the α-cut. When the meaningful set of membership grades are at or above the α-cut (i.e., $\mu_A(x) \geq \alpha$) it is a *strong* α-cut; when the meaningful grades are only above the α-cut (i.e., $\mu_A(x) > \alpha$) it is weak.

The main reason α-cuts are important is that the set of values above that level describes a power (strength) function that is used in fuzzy system models to decide whether a grade of membership should be treated as zero. This becomes an important consideration when rules are being applied to fuzzy sets and when multiple fuzzy sets are combined. The outcome of a group of IF-THEN rules can change meaningfully depending on whether an α-cut has been defined, at what grade of membership it is placed, and whether the cut is weak or strong. Examples of the difference made by setting alpha-levels are presented in Section 9.10 on page 99.

9.5.6 Fuzzy Hedges

In everyday language hedges are "reservations and qualifications in one's speech so as to avoid committing one's self to anything definite."[2] Applied to fuzzy sets, hedges are modifiers that concentrate or dilute the support range of the fuzzy set (Figure 9.12). In other words,

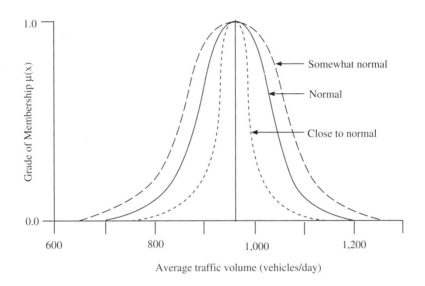

Fig. 9.12. Fuzzy hedges modifying the fuzzy set of *Normal* traffic volume

the hedge changes the shape of the membership curve so that support set values have a different range for inclusion in the fuzzy set. Hedges have two principal effects on the shape of the membership curve, but

---

[2] *Webster's Revised Unabridged Dictionary* (1913). Definition 3 for "hedge" as a verb.

the meaning and interpretation of a hedge is less precise. For environmental impact assessments, this is a benefit that permits adjustment to local values.

Hedges can be concentrators or dilutors. Concentrators are terms such as *very, highly, extremely* that mathematically can be expressed as exponents of the fuzzy membership value. For example, if grade of membership in the fuzzy set *large* is $\mu = [0.9]$, then its membership in the set *very large* is $\mu = [0.81]$ because *very* is conventionally defined as $\mu(x)^2$. Dilutors, on the other hand, are language modifiers such as *slightly, somewhat* and similar terms that change the membership value by a fractional exponent (*e.g.,* $\mu(x)^{1/2}$. A table with hedges and their meaning illustrates the range of language that can be accommodated in designing fuzzy sets for environmental impact assessments (Table 9.1).

Table 9.1. Linguistic hedges and their meaning when applied to fuzzy sets. Adapted from [9, page 218].

Hedge	Meaning
About, around, near, roughly	Approximates a crisp value
Not	Negate or complement
Above, more than	Restricts a fuzzy region
Below, less than	Restricts a fuzzy region
Almost, definitely, positively	Concentrates
Neighboring, close to	Concentrates
Very, extremely	Concentrates
Vicinity of	Dilutor
Generally, usually	Dilutor
Quite, rather, somewhat	Dilutor

The above mathematical definitions of *very* and *somewhat* come from Lotfi Zadeh's early works (e.g., [43]). However, there is no rigorous mathematical reason for using these definitions. They work well in many instances, and when the membership curves are plotted, they look like how most people understand those modifiers applied to a term. In most environmental impact assessments these definitions work very well. If there are reasons to change them to a different definition that better represents the semantics for the preparers or stake-holders, then different definitions will work just as well.

The practical advantages of such linguistic hedges to modify fuzzy membership values is that multiple values (from different stakeholder groups or different experts) can be incorporated into the analysis. One expert might evaluate a wetland's quality as *good* while another ex-pert considers the quality to be *very good*. They are both correct and both opinions are incorporated into the final determination of wet-land quality. Incorporating different professional judgments into the approximate reasoning model does much to reduce—or eliminate—the problem of dueling experts. When appropriate to the assessment, totally contradictory and unconditional rules can be incorporated into the fuzzy rule inference process. The α-cuts described above also play an important role in the determination of the resultant fuzzy set when both conditional and unconditional rules are combined.

Fuzzy logic, probability, and randomness are not the same. The first expresses the subjective nature of linguistic concepts, the second is a measure of uncertainty about the future, and the third is an objective measure—a statistic—of natural systems. Fuzzy sets (the units upon which fuzzy logic operates) represent a belief in the truth of the degree of membership of a measured variable or a concept in one or more linguistic variables.

9.6 Fuzzy Sets (Type-2)

There are uncertainties associated with fuzzy sets that are independent of set type or method of expression. Klir and Wierman[16] recognize three uncertainties: fuzziness, nonspecificity, and strife. Fuzziness (or vagueness) arises from the imprecise boundaries of fuzzy sets. non-specificity (or information-based imprecision) is created by the cardi-nality[3] of a fuzzy set. Strife (or discord) results from conflicts among alternative fuzzy sets. In Type-1 fuzzy sets these uncertainties are all

---

[3] Mathematically, cardinality is the sum of all degrees of membership in the fuzzy set. For example, within a specified assessment area several measure-ments of ground slope are made. The degrees of membership (μ) for each measurement in the fuzzy set, *Steep_slope*, is [1.0], [0.7], [0.8], [0.9], [0.6]. The sum of these values, 4, is the cardinality of the fuzzy set *Steep_slope*.

associated with the position of the membership function along the universe of domain. When the fuzzy set and its membership function are used to model a word that has no underlying measurement, the uncertainties can be considered to be *linguistic* uncertainties [21]. With the appropriate modeling of these uncertainties, the concept of randomness can also be described and evaluated. Type-2 fuzzy sets are more appropriate for modeling the randomness of fuzzy uncertainty than are Type-1 fuzzy sets.

A major category of uncertainty in environmental impact assessments is prediction of future environments under each of the alternatives being evaluated. These predictions are generated by a fuzzy expert system (Chapter 11 on page 129) built using IF-THEN rules (Section 9.8 on page 95). Human knowledge captured in fuzzy environmental impact assessment model rules is uncertain because the words can mean different things to different people; the consequent—THEN part—may differ among a group of experts (they do not agree on what happens when the antecedent—IF part—is true); or noisy data was used to create the rules. While all these uncertainties in meaning and interpretation cannot be fully captured by Type-1 fuzzy sets, they can be fully captured by Type-2 fuzzy sets. Therefore, environmental impact assessments are fully described by a combination of both Type-1 and Type-2 fuzzy sets and by fuzzy numbers (Section 9.5.3 on page 76). Type-2 fuzzy sets are essential to evaluating significance, sustainability, and other societal values.

Almost all development of Type-2 fuzzy sets has been conducted over the past decade by Jerry M. Mendel, his students, and colleagues (see, e.g., [20–22]). It is from this research that application of Type-2 fuzzy sets to environmental impact assessments has been derived.

The types of imprecision and inherent uncertainty completely addressed by Type-1 fuzzy sets are based on measured numbers. That is, Type-1 linguistic variables are imprecise and inherently uncertain expressions of measurements. For example, "steep" slopes, "large" population size, "long" distance, and "high" dissolved oxygen concentration are all linguistic variables based on measurable quantities. As shown in Section 9.5 on page 68, fuzzy sets are the only way to quantify the degree of membership in a linguistic variable of a measured value. These Type-1 fuzzy sets are integral to assessing environmental impacts without the subjectivity of the traditional approach. However, they are not complete descriptors when the values to be quantified are words that cannot directly be measured. Such unmeasurable variables are central to environmental impact assessments; e.g., the evaluation of "significance," "acceptable," "sustainable," "healthy," and "risk." Such linguistic terms also appear in many successful fuzzy system models for business and industry. These business- and industrial-process models have used techniques such as scalable monotonic chaining ([9]) to reach conclusions about words for which there is no underlying measurement.

A Type-2 fuzzy set removes the requirement that the grade of membership be a crisp value by making the membership function fuzzy [43, 45]. A Type-2 fuzzy set grade of membership would be expressed as *about [0.48]* rather than as *[0.48]* without the qualifier. This qualification of membership value offers a more robust way of expressing values associated with words. Another perspective on the differences between Type-1 and -2 fuzzy sets comes from consideration of the different kinds of uncertainty involved in modeling the real world and understanding where randomness fits into the picture. Type-2 fuzzy sets are more mathematically defined by [20] as fuzzy sets whose grades of membership are Type-1 fuzzy sets in the range [0,1]. Extending this line of thinking, there is no particular reason why the grades of membership of a Type- 2 fuzzy set must be sharply defined either. Therefore, there is a Type-m fuzzy set where the membership grades are themselves Type- m fuzzy sets, with $m > 1$, and each one of the m fuzzy sets having values in the range [0,1]. In reality, these higher-order fuzzy sets are both more difficult to apply semantically to real world problems and computationally highly difficult. The important point is that it is possible mathematically to express highly uncertain, imprecise, or subjective concepts regardless of how abstract they are.

To illustrate the differences between Type-1 and Type-2 fuzzy sets compare Figure 9.4 on page 74 with Figure 9.13 on the next page. Both representations of the linguistic variable *Significant* use an arbitrary scale of values for the x-axis, but represent the grade of significance quite differently. As a Type-1 fuzzy set for *Significant*, the grade of membership can be established for any point along the x-axis by raising a line perpendicular to that axis until the membership function is reached, rotating the line horizontally to the left until it intersects the y-axis and reading (or interpolating) the grade of membership from the y-axis. This procedure works effectively when there is general agreement on the membership function shape and its expression of the un-

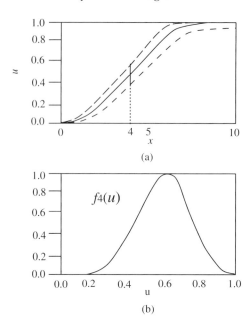

Fig. 9.13. (a) Graphical representation of a Type-2 fuzzy set for *Significance*. The domain of the grade of membership associated with x = 4 is shown by the vertical line. (b) The Gaussian secondary membership function at x = 4.

derlying concept. When such agreement does not exist, or when there is a desire to model, in a more abstract way, concepts that are not expressions of measurements, a Type-2 fuzzy set membership function is the appropriate representation. The upper plot in Figure 9.13, *a*, represents *Significant* on a scale of [0,10]. It is an S-shape curve rather than a shouldered, straight-line plot as in the Type-1 fuzzy set. More importantly, it has a domain associated with each value along the x-axis. This domain of uncertainty is neither constant from left to right, nor necessarily of equal width on both sides of the solid curve. The upper bound of uncertainty cannot exceed the value [1.0], but the lower bound remains less than [1.0]. To illustrate the difference between Type-1 and Type-2 fuzzy sets for *Significant*, note that with the latter type of fuzzy set the value, x = 4 has a grade of membership in the range [0.38—0.55]. The lower part of the figure, *(b)*, shows the distribution of values within that membership grade. Think of this lower curve as having been turned 90 degrees from its normal position perpendicular to the plane of the page. The highest grade of membership in the secondary membership function is about [0.6]. This means there is complete confidence that [0.48] is the grade of membership of *Significant* when x = 4. Confidence in grades of membership decreases from that value toward both the upper and lower bounds.

Type-2 fuzzy sets can be difficult to comprehend when first encountered. The need to describe linguistic variables based on unmeasurable concepts is easy enough to grasp. In fact, it makes so much sense it is difficult to believe it is not applied more widely. But visualizing how the concept is implemented is not intuitive. Figure 9.14 presents a sim-

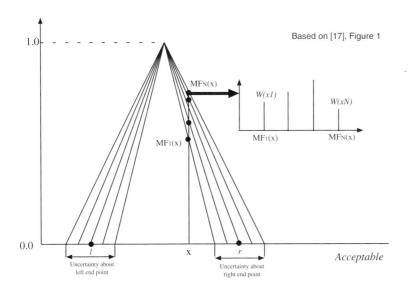

Fig. 9.14. A triangular fuzzy set for the concept of *Acceptable*. The base points of the member function (marked *l* and *r*) span a range of values

ple Type-2 fuzzy set expressed by a triangular membership function. At each value of the *primary variable* (x in the figure is one such value of the primary variable) the grade of membership is located within a *secondary membership function* (a vertical slice perpendicular to the surface of the page) rather than a point value. That is, rather than crisp values for the grades of membership there is a *footprint of uncertainty*

(FOU) associated with the membership function curve. This secondary membership function is represented by points defining its shape; this is illustrated in the upper, right corner of Figure 9.14 on the page before. This *secondary grade of membership* is the heights of the lines in the inset figure. As with all fuzzy sets, values within this secondary grade of membership are in the interval [0,1]. It is this new, third dimension that accommodates uncertainties in the value of the grade of membership associated with any given value along the x-axis. To complete this introduction, the Universe of Discourse (or domain) for this secondary membership function ($MF_1(x)$ to $MF_N(X)$) is the *primary membership function* and is in the interval [0,1].

As with Type-1 fuzzy sets, Type-2 fuzzy sets can have different shapes. The Gaussian curve is one of the most commonly applied and it has two distinct forms. In addition, Type-2 fuzzy sets can be continuous or interval; the former are rarely used in practical applications so discussion is limited to the latter type. Type-2 fuzzy sets defined by Gaussian curves can have uncertain means, uncertain deviations, or both.

A Gaussian curve with an uncertain mean (Figure 9.15 on the facing page) has a footprint of uncertainty that varies in magnitude based on the values of both the primary variable (the position along the x-axis) and the primary membership (the Universe of Discourse; range of the x-axis). This behavior reflects increased uncertainty in meaning of terms such as *Significant* as values diverge from the center of the range.

The upper bound of the uncertain mean curve is truncated at $u = 1$ because no grade of membership can exceed [1.0]. The primary membership function has a fixed standard deviation, σ, and an uncertain mean that takes values within the closed interval $[m, m_2]$; that is,

$$\mu_\Gamma(x) = exp\left[-\frac{1}{2}\left(\frac{x-m}{\sigma}\right)^2\right] \quad m \in [m_i, m_2] \qquad (9.5)$$

This primary membership function is very well suited to characterize the antecedents and consequents in approximate reasoning (IF-THEN) models where the rules involve unmeasurable concepts or words. Such is the case in the impact assessment stage of an environmental impact assessment because there is no way to accurately position the mean value of the curve.

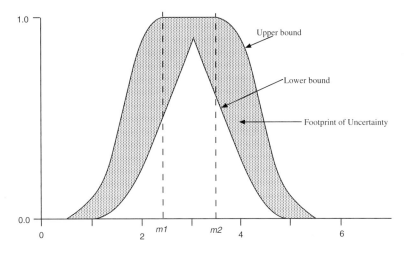

Fig. 9.15. A Type-2 fuzzy set as a Gaussian curve with uncertain mean.

The other type of Gaussian primary membership function has a fixed mean, m, and an uncertain standard deviation with values in the interval $[\sigma_1, \sigma_2]$ (Figure 9.16 on the next page); that is,

$$\mu_\Upsilon(x) = exp\left[-\frac{1}{2}\left(\frac{x-m}{\sigma}\right)^2\right] \quad \sigma \in [\sigma_1, \sigma_2] \tag{9.6}$$

The uncertain standard deviation primary membership function is used to model measurements corrupted by nonstationary, additive noise. This situation may occur during the environmental characterization stage of an environmental impact assessment if data are being analyzed by time series to predict a future state.

The following sections will be restricted to operations on Type-1 fuzzy sets. When operators, implication methods, defuzzification, and the like are to be applied to Type-2 fuzzy sets, the differences will be explained in the context of a process specifically related to an environmental impact assessment stage. References cited above in this section provide comprehensive coverage of the theory and practice of Type-2 fuzzy sets for those who want to delve more deeply into the subject.

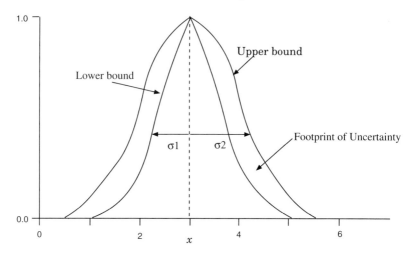

Fig. 9.16. Gaussian primary membership function with uncertain standard deviation. The FOU disappears to certainty when $\mu_\Upsilon(x) = 1.0$.

9.7 Fuzzy Set Operators

Fuzzy logic is a superset of Boolean logic and is based on the equivalent of Boolean truth tables. These truth tables determine the results of the logical operations AND, OR, and NOT values that are true (1) and false (0) (Table 9.2). Because the grade of membership within a

Table 9.2. Boolean truth tables for *AND, OR,* and *NOT.*

$$AND \begin{array}{c|cc} & 0 & 1 \\ \hline 0 & 0 & 0 \\ 1 & 0 & 1 \end{array} \qquad OR \begin{array}{c|cc} & 0 & 1 \\ \hline 0 & 0 & 1 \\ 1 & 1 & 1 \end{array} \qquad NOT \begin{array}{c|c} 0 & 1 \\ \hline 1 & 0 \end{array}$$

fuzzy set can be any value between [0,1] the dichotomous Boolean truth tables are replaced by functions that produce the same results (Figure 9.17 on the next page). The set operation intersection or conjunction (∩) results in a new set with all values common to both sets. It is the minimum value of the two fuzzy sets and is written as *min(A,B)*

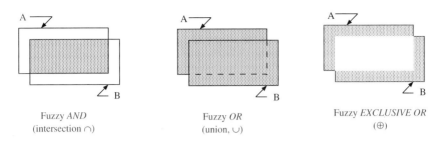

Fuzzy *AND*
(intersection ∩)

Fuzzy *OR*
(union, ∪)

Fuzzy *EXCLUSIVE OR*
(⊕)

Fig. 9.17. Fuzzy set operators *min, max,* and *complement.*

or $\vec{A} \wedge \vec{B}$. The set operation union or disjunction (∪) results in a new set with all values of both sets. It is the maximum value of the two fuzzy sets and is written as *max(A,B)* or $\vec{A} \vee \vec{B}$. The fuzzy set equivalent of *NOT* is the complement of the fuzzy set, defined as $\neg \vec{A} = \overrightarrow{\neg A}$ or $\neg \vec{A}$. These definitions of how to combine fuzzy sets were first described by Lotfi Zadeh [42].

Fuzzy set intersections are mapped by a general function, T, which combines the two functions as $\mu_{A \gg B}(x) = T(\mu_A(x), \mu_B(x))$. All the binary operators that intersect two fuzzy sets are referred to as *T-norms*, or triangular norms, because they find the minimum value where two fuzzy sets intersect (Figure 9.18). Fuzzy set unions are mapped by a general function, S, which combines the two functions as $\mu_{A \gg B}(x) = S(\mu_A(x), \mu_B(x))$. The binary operators that union two fuzzy sets are referred to as a *T-conorm*, or *S-norm*, because they calculate the maximum value of the union of two fuzzy sets (Figure 9.19 on the following page). There are many more fuzzy operators than just min and max. Details and rationales can be found in [9, 45].

Using fuzzy set operators in the performance of an environmental impact assessment frequently requires the use of a combination of AND and OR. This combination makes sense because AND is too restrictive, while OR is not sufficiently restrictive. Consider a jurisdiction that statutorily declares an aggregate resource (sand, gravel or rock) to be significant only if the resource size estimated by qualified geologists is at least 2,000,000 pounds, the hardness and wear meet road construction standards, and the soils overlaying the deposit are not considered prime farmland soils. Within this jurisdiction a company

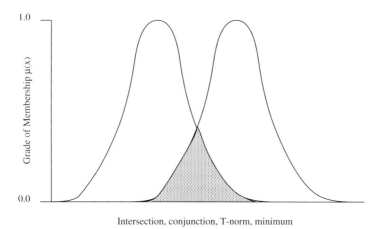

Intersection, conjunction, T-norm, minimum

Fig. 9.18. Fuzzy *AND.*

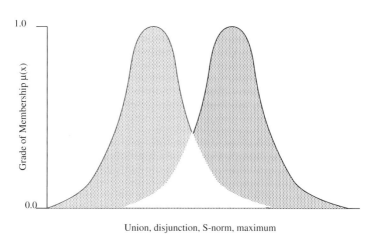

Union, disjunction, S-norm, maximum

Fig. 9.19. The union of two sets, *A* and *B*, and the T-conorm, or S-norm.

identifies a property containing an estimated 1,975,000 pounds of aggregate that far exceeds quality thresholds throughout the depth of 65 feet on nonprime farmland. Because the definition of "significant aggregate resource" requires meeting criteria of quantity AND quality AND location this resource is not available for use in constructing roads or buildings. The crisp quantity threshold has not been met and the increased quality and location values cannot compensate for the just-too-low quantity value. The opposite can occur when a set of criteria are joined by OR. In this case only one of the criteria needs to be fulfilled to qualify. In many cases that is not the intent when defining criteria, so what is really needed is a combination of AND and OR. The Ordered Weighted Average (OWA) is one family of compensating operators that have varying degrees of ANDness or ORness. These operators can take into account the importance of the information source (e.g. the qualifications of the expert), the relative values of the different criteria and other weights and priorities of the decision-makers of society in general.

Completing an environmental impact assessment that is technically sound and reflective of society's values and concerns requires that alternative, compensatory, forms of the fuzzy intersection and complement operators be used. The fuzzy union operator is used less frequently, but compensatory forms provide a much better match to reality than does $max(A,B)$. In several stages of the EIA process the ability to have an operator take on the properties of both AND and OR means that accurate expression of criteria can be expressed and calculated. Additional insight into the use of compensatory operators is provided in the next two sections as well as the chapters describing the application of fuzzy logic to stages of an environmental impact assessment.

9.8 IF-THEN Rules

Linguistic variables and their associated fuzzy term sets are the data upon which fuzzy logic is applied. The operators are the mechanism that allow the sets to be combined. The IF-THEN rule statements describe the conditions for combining fuzzy sets. A simple fuzzy IF-THEN rule assumes the form

```
IF x is A THEN y is B
```

where A and B are linguistic variables defined by fuzzy term sets on the ranges Universes of Discourse) X and Y, respectively. The IF-part of the rule (x is A) is called the *antecedent* or *premise*, while the THEN-part of the rule (y is B) is called the *consequent* or *conclusion*. An example of such a rule might be—

```
IF Rainfall is heavy
   THEN Erosion_risk is increased.
```

The antecedent to this rule is a single value: the grade of membership of the rainfall in the fuzzy set "heavy" (for example, $\mu_{heavy} = [0.4]$) while the consequent is a complete fuzzy set, "increased." Frequently the the IF-THEN rule contains multiple antecedents that are applied to the same consequent fuzzy set. A more accurate representation of the above example would be

```
IF Rainfall is heavy
   AND Slope is steep
   AND Vegetation_cover is low
   AND Soil_erodability is high
      THEN Erosion_risk is greatly increased.
```

The grade of membership values for each fuzzy set in the antecedents will be combined by an AND or compensatory-AND operator to produce the consequent fuzzy set. This resultant set can be *defuzzified* by one of several methods to yield a crisp value for the calculated grade of membership in the hedged fuzzy set, "greatly increased." The above rule is illustrated in Figure 9.20. Although the slope has a low grade of membership in the fuzzy set *Steep* and the rainfall is not extreme, the very high grade of membership in the variable "Soil_erodability" resulted in that value's being assigned to the consequent of greatly increased erosion risk. Such a domination of a single factor in a multi-factor antecedent may be appropriate in some environmental impact assessments but certainly not in all of them. This is where the judgment of the practitioner as a knowledge engineer must be applied to selecting the most appropriate fuzzy operator for a given situation.

What is not always obvious in Figure 9.20 is that the three antecedent rules are processed in parallel, not sequentially. In a parallel-processing fuzzy modeling system, the consequent fuzzy set is not available for use in further processing before all the antecedent conditions have been applied to it. The results could be quite different when

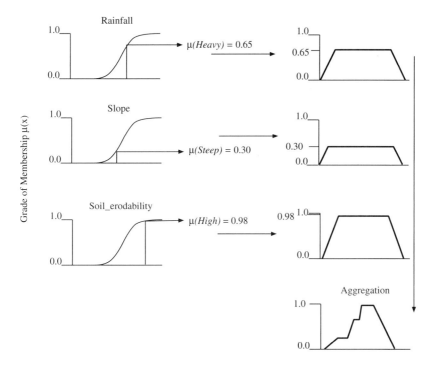

Fig. 9.20. Antecedent conditions combined with *AND* using the fuzzy operator *max(A,B)*.

rules are processed serially rather than in parallel. With serial process-ing of antecedent rules the order in which they are listed can affect the value of the consequent fuzzy set. What is needed, of course, is con-sistent output; that is, output that reflects each of the input fuzzy sets equally and appropriately.

Most fuzzy models describe large and highly complex systems. En-vironmental impact assessments are included in the class of large and highly complex systems to be modeled. The rules used to character-ize the baseline conditions and predict future conditions based on the alternatives presented can number in the dozens or hundreds. With-out an organizing scheme a model can become unwieldy to create and maintain. The solution to organize the complexity is the concept of *poli-cies*.

A policy is a logical unit of the overall model and may consist of nested subpolicies. Each subpolicy is evaluated as a distinct processing unit before being appropriately combined with other sub-policies to create a consequent fuzzy set for the overall policy. As an example, consider the baseline condition components shown in Section 4.1. One way of organizing this model would be with four policies (Ecology, Aesthetics, Physical/Chemical, and Human Interest/Social), each of which has several subpolicies and sub-subpolicies. The Ecology policy would have subpolicies for terrestrial species and populations, aquatic species and populations and terrestrial habitats and communities. The Human policy would have subpolicies for education/scientific and historical. The latter might be further subdivided to capture all the important input data.

9.9 Defuzzification

Consequent fuzzy sets need to be converted back to a crisp number. These resultant numbers represent the existing conditions and the projected future conditions based on the alternatives. The process of representing a consequent fuzzy set as a crisp number is called *defuzzification*. The value of the consequent set can be determined by several methods. The two most commonly used in expert and decision support systems are use the center of gravity and the maximum of output (Figure 9.21). While many approaches to finding a single number rep-

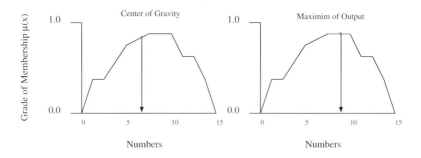

Fig. 9.21. Two methods of defuzzifying a consequent fuzzy set.

resentative of the result fuzzy set have been proposed, none of them has a fundamental set-theoretical basis. However, both the center of gravity and maximum of output methods are mathematically based (even if not derived from fuzzy set theory) and are suitable for the semantics inherent in an environmental impact assessment. Professional judgment and experience are the best guide for which one to use in a particular model, as the two methods produce different crisp numbers. Giving consideration to all factors, the center of gravity method makes a good default that will be appropriate most of the time.

The center-of-gravity method combines evidence from all antecedent rules, weighing the result based on the relative truth of each rule. It is based on Bayesian probability and is the equivalent of the weighted average of the consequent fuzzy set. The crisp result, R, is calculated using this equation,

$$R = \frac{\sum\limits_{i=o}^{n} d_i x \mu_i}{\sum\limits_{i=0}^{N} \mu_i} \tag{9.7}$$

where d_i is the domain value and μ_i is the grade of membership for that domain point. The maximum of output method uses the domain value associated with the highest grade of membership (if it is a single point), or the midpoint of a plateau at the greatest grade of membership.

While the lack of proven, axiomatic basis for these defuzzification may at first appear to be a deficiency in the rigor of fuzzy system modeling, it has been proven by years of experience in the application of such models for control, decision support, risk analysis, fraud detection, and other complex models. Also, in conventional (i.e., nonfuzzy) expert system or decision support models the results are produced by interaction of many more rules with discrete upper and lower limits by methods that are usually less well understood than is the case with fuzzy system models. The bottom line is that they work.

9.10 Fuzzy Implication

Approximate (and plausible) reasoning permit conclusions to be drawn from hypotheses. In classical logic there are several modes of reasoning (e.g., *modus ponens*[4] and *modus tollens*[5]), tautologies[6] that allow a conclusion to be drawn from predicates. If we relax the classical requirement that statements be crisp values and, instead, permit statements to contain fuzzy sets of linguistic variables then we are using approximate reasoning. These modes of reasoning are called implication methods because the truth of one side of an equation determines the truth of the other side of the equation. That is, one truth value implies the other. These implication rules are implemented within a fuzzy system model by the fuzzy inference engine.

Modus tollensstates that "If A, then B. B is false. Therefore, A is false." In the set-theoretic notation applicable to fuzzy logic this is written

$$A \subseteq B \tag{9.8}$$
$$x \notin B \tag{9.9}$$
$$\therefore x \notin A \tag{9.10}$$

and read as, "A is a subset of B; x is not in B; therefore, x is not in A." Modus tollens states that if the consequent of a conditional is false, then the antecedent must also be false. Consider, "If it rained last night, then the grass would be wet. The grass is dry. Therefore, it did not rain last night." Modus ponens is the opposite of modus tollens. That is, "If A, then B. A. Therefore, B." In set theoretic notation this is written—

$$A \rightarrow B \tag{9.11}$$
$$\vdash A \tag{9.12}$$
$$\vdash B \tag{9.13}$$

where the symbol, $\vdash$, represents "logical assertion." An example of modus ponens is, "If salmon redds are present, this must be a spawning stream. Salmon redds are present. Therefore, this is a spawning stream." Both modus ponens and modus tollens are used in fuzzy inference systems applied to the IF-THEN rules and provide the basis for confidence in the conclusions of the inference engine.

---

[4] Latin: mode that affirms.
[5] Latin: mode that denies.
[6] A proposition, or statement, that is always true.

Implication methods are the mathematical means for transferring the truth of antecedent fuzzy sets to the resultant fuzzy set. The fuzzy system models for environmental impact assessments have multiple antecedent rules implying a single output set. These multiple input rules are *aggregated* to produce the final output (consequent) fuzzy set as illustrated in Figure 9.20 on page 97. There are also instances in which a single antecedent fuzzy set can transfer its truth to multiple consequent fuzzy sets. For example—

```
IF Land_use is range
   AND Vegetation_cover is very low
      THEN Wildlife_habitat is decreased
         AND Soil_erosion is slightly increased.
```

There are many implication methods used in fuzzy expert and decision support systems. Those most commonly used in environmental impact assessments include *min-max, additive, geometric mean, symmetric summation,* and *ordered weighted average.*

9.10.1 Min-Max Implication Rules

The *min-max* rules apply the minimum truth of each antecedent to the consequent fuzzy set. Mathematically this is expressed by two equations. The first equation,

$$\mu_{cons}[x_i] \leftarrow min(\mu_{pred}, \mu_{cons}[x_i]) \tag{9.14}$$

declares that the consequent fuzzy set is assigned the value of the *minimum* of the truth value of the antecedent fuzzy set or its current truth value. The solution fuzzy set is updated by the maximum of the consequent fuzzy sets according to the second equation,

$$\mu_{soln}[x_i] \leftarrow max(\mu_{cons}[x_i], \mu_{soln}) \tag{9.15}$$

This two-step process is illustrated in Figure 9.22 on page 103. when a value has a partial membership in both sets, the minimum value is used. The first rule,

```
IF Rainfall is heavy,
```

evaluates to a truth of [0.25] because the rainfall of 0.95 cm/hr has a much higher truth ([0.70]) in the fuzzy set "moderate" (only these two fuzzy sets are shown from the fuzzy term set for Rainfall. The second rule,

```
IF Slope is steep,
```

has a truth value of [0.10] because, as the term set was defined for this hypothetical project, a slope of 17.5 percent has a much higher grade of membership in the fuzzy set "very_steep" ([0.65]). The third rule evaluates the truth of

```
IF Vegetation_cover is low,
```

and this is true to the extent [0.45]. For this rule there is only one fuzzy set that is associated with the measured value of 15 percent vegetation cover. Therefore, this single value is transferred to the consequent fuzzy set. The last rule in this policy,

```
IF Soil_erodability is high,
```

has a truth of [0.48] (the grade of membership in the fuzzy set "high").
When minimum values have been transferred to the consequent fuzzy set for all the predicate variables, the maximum of those minimums is the value contributed to the solution fuzzy set. If the rules in the policy for slope stability were part of an evaluation of land suitability for timber harvest, the grade of membership, $\mu_{Erosion-risk} = [0.48]$, would be aggregated with the results of other policies in the consequent fuzzy set, Land_suitability. Alternatively, the defuzzified value of the solution set (shown by the position of the arrowhead on the x-axis) can be used.

9.10.2 Additive Implication Rules

The additive implication rules are similar to the min-max rules in the generation of the consequent fuzzy sets but differ in the generation of the solution fuzzy set. The consequent fuzzy set is still created by the minimum of the current truth value or the predicate set truth value,

$$\mu_{cons}[x_i] \leftarrow min(\mu_{pred}, \mu_{cons}[x_i]) \qquad (9.16)$$

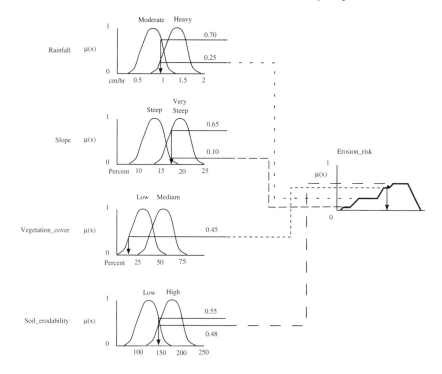

Fig. 9.22. Applying the rules using the min-max inference method.

but the solution fuzzy set is updated by a different rule,

$$\mu_{xoln}[x_i] \leftarrow (\mu_{soln} + \mu_{cons}[x_i]) \qquad (9.17)$$

which adds the truth value of the consequent fuzzy set to the accumulating solution fuzzy set. The maximum value of μ_{xoln} can exceed [1.0] using the additive implication rules. If the centroid defuzzification method is applied, then this does not matter. If another defuzzification method is used (e.g., maximum of output) then the solution fuzzy set needs to be *normalized;* that is, scaled so the maximum value if [1.0] and all other values are reduced proportionally. There is another operation sometimes used in fuzzy control systems called the "bounded sum operation." The equation for this operation is that of the additive implication rule for determining the solution fuzzy set.

However, the bounded set operation truncates the maximum value at [1.0] rather than allowing it to exceed the normal grade of membership (truth value). It is clear that truncating the solution fuzzy set can result in loss of valuable information that is retained by using the additive rules then normalizing the solution fuzzy set when necessary.

9.10.3 Weighted Geometric Mean

The weighted geometric mean is similar to the arithmetic mean (average) but calculated as the nth root of the product of the individual measured values multiplied by their relative weights of importance—

$$\Pi((a_1 w_1) \cdot (a_2 w_2) \cdot (a_3 w_3) \cdot (a_4 w_4) \cdot .... \cdot (a_n w_n))^{1/n} \qquad (9.18)$$

.

The main advantage of using the geometric mean over the arithmetic mean when conducting an environmental impact assessment is the reduction in influence in the highest and lowest values. In approximate reasoning models of subjective societal values (such as the relative importance of different environmental conditions), decreasing the influence of the extremes on the mean value produces a more realistic statistic.

As shown above, the values being averaged are weighted for their relative importance. One group of stakeholders may consider wildlife habitats to be the most important value of an environment, so they will give the wildlife habitat membership value a high weight. Another group of stakeholders may give wildlife habitat a lower weight because, to this latter group, economic value has the greatest importance. Pragmatically this approach permits the inclusion of a range of beliefs and values with appropriate weight scaling for each group. All values are used in the calculation of a final result.

9.10.4 Symmetric Summation

As noted in Section 9.7 on page 92, the AND and OR operators are quite restrictive: the worse case determines the outcome. With the AND operator, every criterion must be true for the combination of two sets to be accepted. With the OR operator, only a single criterion need be true for the combination to be accepted. While the need to compensate for all the in-between cases was addressed in that section, nothing was presented for the case of a set and its complement, the NOT operator. This important issue is seen whenever some consideration must be made for a value being partially "good" and partially "bad"—or, somewhere between acceptable and unacceptable. Within the broad range of applied environmental decisions, decisions based on a single criterion ranging from "good" to "not good" are very common.

A good environmental example of choosing between "Good" and "NOT Good" is presented by Lake Tahoe, on the east side of the Sierra Nevada Mountains and astride the border of California and Nevada. Until the mid-1800s the lake was blue, crystal clear, and biologically sterile. After silver deposits were discovered in the Comstock Lode just north of the lake (centered on Virginia City, Nevada), the forests along the west side of the lake were heavily logged. The wood was used as mine-support timbers, for houses and other buildings, and for fuel. As slopes were deforested, soils and nutrients washed into Lake Tahoe by precipitation and snowmelt runoff. These nutrients allowed a biological food web to form with microscopic plants (phytoplankton) supported by the nutrient input from the drainage basin, microscopic animals (zooplankton) feeding on the plants, and so on up the food web until there developed a large, commercial fishery for trout. The fish fed local populations and were shipped to San Francisco, too. During the past decade, a powerful movement has developed to "save Lake Tahoe" and return it to its former blue, clear state where objects can be seen very deep below the surface. This is the "Good"; a socially supported desire to return the lake's waters to the color and clarity they once had. The "NOT Good" side of the equation is that only by killing off the biological productivity of the lake can this physical change be accomplished. This means no fish, no boating and possibly no swimming or other recreational activities that would put nutrients into the water. This situation can be decided by an either-or choice of extremes or some middle ground. The resolution of how to address this type of situation was provided by Silvert [31] when he showed that a set and its complement can be combined in a useful way.

The definition of the set *NOT A* is: $\neg \vec{A} = 1 - \vec{A}$. If any combination of two fuzzy sets is described as $\vec{A_1} \bowtie \vec{A_2}$, then it is also true for their complements

$$\neg(\overrightarrow{A_1} \bowtie \overrightarrow{A_2}) = \neg\overrightarrow{A_1} \bowtie \neg\overrightarrow{A_2} \tag{9.19}$$

that is, the complement of two fuzzy sets joined by an operator is equal to the complement of each fuzzy set joined by the same operator. The many rules for combining two fuzzy sets (intersection, union, bounded sum, etc.) do not comply with the symmetry condition defined in equation 9.19.

Silvert developed an operation that satisfies the symmetric combination condition. His combination rule is valid for any nonnegative function $g(\mu_1, \mu_2)$:

$$\mathbb{C}(\mu_1, \mu_2) = \frac{g(\mu_1, \mu_2)}{g(\mu_1, \mu_2) + g(1 - \mu_1, 1 - \mu_2)} \tag{9.20}$$

by generating a symmetric sum for a fuzzy set and its complement. The key to producing practical and meaningful results is the choice of the function, $g(\mu_1, \mu_2)$; this condition is fulfilled when $\mathbb{C}(\mu, 1 - \mu) = \frac{1}{2}$.

Symmetric summation has important applications in the conduct of environmental impact assessments; for example, in characterizing baseline conditions and the impacted environments predicted for the alternatives considered.

9.10.5 Ordered Weighted Aggregators

This section began with a explanation of the problems associated with the excluded middle. The two extremes are that all criteria must be fulfilled before the consequent truth can be accepted (AND) or only one criterion need be fulfilled for the consequent truth to be accepted (OR). When conducting an environmental impact assessment, it is the middle ground that applies: that is, some criteria with a high truth value can compensate for other criteria with low truth value to derive a value between the two extremes. Also, many assessments assign different weights to each criterion; they are all of different value to society and the decision-makers. The explanation of multi-objective, multi-criteria decision-making in Section 1.1.1 on page 4 needs a fuzzy implication method to allow such decisions to be made. The symmetric summation method meets some of these needs, but in the more restrictive case of a fuzzy set and its complement (NOT).

Yager [41] successfully addressed these problems by developing a class of fuzzy set aggregators that have adjustable "orand" capabilities and apply to multiple criteria with either equal or different importance weights. These aggregators are called *ordered weighted averaging (OWA)* operators and they fall between the extremes of the t-norm (ANDing) operators and the t-conorm (ORing) operators. Yager's definition of an OWA is a mapping (F) from $I^n \rightarrow I$ (where $I = [0,1]$) if there is a weighting vector W associated with the mapping function F. This weighting vector has the general form

$$W = \begin{bmatrix} W_1 \\ W_2 \\ W_3 \end{bmatrix} \tag{9.21}$$

and fulfills two conditions—

1. $W_i \in (0, 1)$
2. $\sum_i W_i = 1$

The weights, W_i are associated with a specific ordered position, not a particular element. An example will make clear this distinction. Let F be an OWA operator of size $n = 4$ and the weight vector, W, be

$$W = \begin{bmatrix} 0.2 \\ 0.3 \\ 0.1 \\ 0.4 \end{bmatrix} \tag{9.22}$$

Then calculate $F(0.6, 1.0, 0.3, 0.5)$.
The ordered argument vector is

$$B = \begin{bmatrix} 1.0 \\ 0.6 \\ 0.5 \\ 0.3 \end{bmatrix} \tag{9.23}$$

So, $F(B) = W'B = [0.2, 0.3, 0.1, 0.4] \begin{bmatrix} 1.0 \\ 0.6 \\ 0.5 \\ 0.3 \end{bmatrix} = 0.55 \tag{9.24}$

In an environmental impact assessment, the vector B might be the defuzzified characteristic values of alternative future conditions, while the weighting vector, W, might be the relative values of different criteria to be applied to each alternative. The value of the operator, F, (0.55 in the example above) is then the rating of that alternative and those criteria. decision-makers can then use the set of results to compare among all alternatives and to place the same operator value for the existing conditions in that range [0,1].

9.11 Compatibility Index

Earlier sections in this chapter explained how to manipulate fuzzy values to produce rigorous output when the underlying values or numbers are subjective. The rest of the chapters in this part of the book shows how these components are applied to environmental impact assessments to remove limitations of the traditional approach. Before the modern approach is explained, there is one remaining fuzzy system component that must be understood and properly applied. This tool is used to measure the robustness of the fuzzy system model. That is, it measures how well the output fuzzy set answers the questions being asked. Developed by Earl Cox [8, 9], this tool is called the *compatibility index* (CIX). The compatibility index can be considered the fuzzy equivalent of a goodness-of-fit or confidence interval used with probability-based statistics. The important questions answered by calculation of the CIX include—

- How well do the input data and model logic work together to produce the observed result?
- How consistent is the strength of the model recommendations from one run to another?
- What confidence can we have that the model is properly functioning?

Input (antecedent) fuzzy sets are created by the knowledge engineer and domain experts to capture the underlying semantics of the linguistic variables. However, the fuzzy sets that form the solution are created by the application of the rule base and the aggregation methods used. These consequent fuzzy sets reflect the degree of truth contained in the model and how well the model's rules respond to the model's input data. In other words, the relationships among input data, processing rules and solution set is represented by the compatibility index.

Cox [9] defined two types of model compatibility: statistical and unit. Statistical compatibility is a measure of model performance over a large data range; it is a true measure of system compatibility. Unit compatibility measures the strength of the output recommendation from running a model once. The idea underlying the compatibility index is based on the height of the consequent fuzzy region. If the height is close to 1.0 or 0.0 then the model assumes the Boolean output of the nearest extreme. The solution fuzzy set membership function is very high or very low when the output data are at the extremes of the range and this occurs when the predicate truth of the result is indistinguishable from one or zero. This region of decreased compatibility with the truth of the input data and model is illustrated in Figure 9.23.

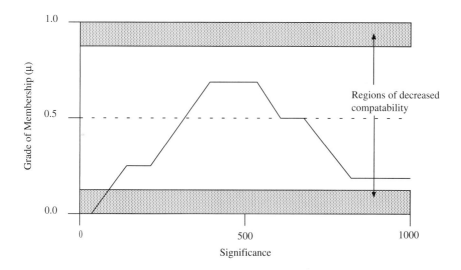

Fig. 9.23. Regions of decreased compatibility in a solution fuzzy set.

9.11.1 Unit Compatibility Index

In every fuzzy system model there is one unit compatibility index per solution variable. The unit compatibility measures the height of the

solution fuzzy set while defuzzification to a crisp output value is independent of the maximum height of the set. In Section 9.9 on page 98
the two most commonly used defuzzification methods were explained.
Regardless of method (composite maximum or center of gravity), the
resulting crisp value depends on the solution set width and shape but
is independent of height.

In an environmental impact assessment, there is a rule for impact
assessment of an alternative that evaluates the significance of erosion
risk under that alternative—

```
IF Rainfall is heavy
   AND Slope is steep
   AND Vegetation_cover is low
   AND Soil_erodability is high
      THEN erosion_risk is highly significant.
```

When the model is run under two sets of input data the defuzzified
crisp output for both data sets is 428.5 (Figures 9.24 and 9.25). It is rea-

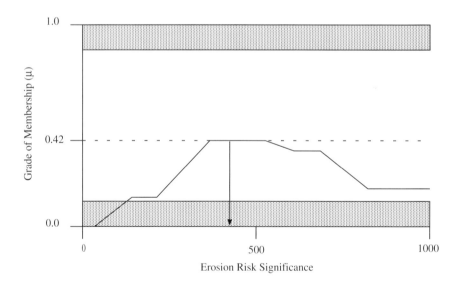

Fig. 9.24. Defuzzified output of erosion risk significance when $\mu[0.42]$.

sonable to ask why two different input sets produce the same value for erosion risk significance even when the rules are the same. The unit compatibility index is the numeric explanation. The first model run generated a solution fuzzy set with a grade of membership—compatibility index—of [0.42], while the second run generated a solution fuzzy set with a CIX of [0.72]. While both produce the same crisp value after defuzzification, the strength of our belief in the truth of this output is greater with the second set of input data. In other words, the model's recommendation that the erosion risk is highly significant is stronger in the second run than in the first run. The recommendation is stronger because the input data fall well within the antecedent fuzzy sets; their truth membership is relatively high. Therefore, the defuzzified output has a higher degree of support in the second run. Quantifying the relative strength of an output value (i.e., its robustness) is an important task in most environmental impact assessments.

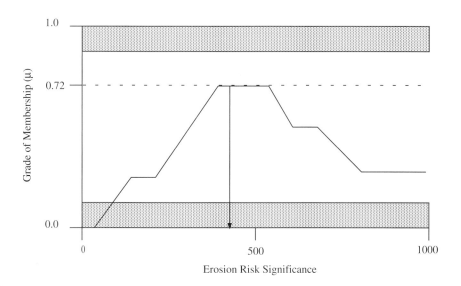

Fig. 9.25. Defuzzified output of erosion risk significance when $\mu[0.72]$.

9.11.2 Statistical Compatibility Index

Generally, an environmental impact assessment requires only a single fuzzy system model run for each alternative. However, with the complex computational work performed by a computer there is the opportunity to explore input options to the alternatives. Because the forecast impacts are based on professional judgments of experts, there are advantages to the decision-makers and stakeholders to running many sets of input data, each representing a slightly different scenario for each alternative. When a series of results have been produced it is necessary to evaluate their relative strengths and gain insight into the fit of the model to support a decision. In these cases, the appropriate measure is the numeric average of each model run's compatibility index over a sufficiently large number of runs. Even the most comprehensive attempts to capture variation in input values will result in comparatively few runs, probably fewer than a half-dozen. This limits the value of the statistical compatibility index for environmental impact assessments. However, in other fuzzy system models, there are opportunities to run the same model hundreds or thousands of times with different input data. With these models the fuzzy concept of *sufficiently large number* of model runs can easily be met.

With a large number of model runs, trends in the CIX can be identified, shifts in input value clusters will produce measurable output changes, and model variability can be accurately determined. Although this opportunity is not present in an environmental impact assessment, there are insights to be gained on the suitability of rules for the expected range of input values.

9.11.3 Compatibility Height Selection

When the defuzzification method uses the center of the maximum value in the solution fuzzy set the compatibility index for the point of defuzzification is the maximum height of the output membership function. But when the defuzzification method is the center of gravity, the height of the membership curve at that point may not be the maximum height. In the latter case, the compatibility index is calculated at the point of defuzzification.

The CIX at the point of defuzzification represents both the minimum working height of the solution fuzzy set and the point at which the crisp output value is calculated. This value is normally between $\mu = [0.40]$ and $\mu[080]$ for a well-constructed rule system. Very high ($\mu > [0.92]$) and very low ($\mu < [0.18]$) compatibility indices indicate problems with the model. Either the antecedent fuzzy sets are not properly constructed or the implication rules and aggregation methods are not adequate descriptions of the important processes involved.

10

Environmental Conditions

10.1 Inadequate Data

No matter the size or importance of a proposed project, there are constraints of time, money, and practicality that can affect how data are collected. Even the most carefully and accurately measured data are limited in meaning and use without metadata[1] to provide their context. For environmental impact assessments, metadata describing space, time, and relative position are necessary for the development of fuzzy sets that accurately reflect the underlying semantics of the model. It is not enough to know the size of all wetlands, the number of tree species, or the number of songbird species identified on the project site. To express these data properly in fuzzy term sets the practitioner must know where these components are located, their spatial relations to each other and other environmental components, when the data were collected, and the relative amounts of each data type. By now, almost everyone understands the importance of spatial and temporal context for baseline conditions. Relative amount is not as well known or appreciated.

Relative amount is a partial answer to the question of the importance or value of an environmental component. Size, distance, and similar measurements gain meaning (e.g., in terms of *significance*; see Section 11.3 on page 149) when they are represented as membership values in one or more fuzzy sets. Is the amount of wildlife habitats on

---

[1] Data about data.

the project site small, moderate, or large; is it *adequate*? Are distances to similar wildlife habitats close or far? Are the values of these habitats considered poor, fair, good, or excellent? Expressing the measured values as fuzzy memberships gives them more easily understood values and allows for considerable analyses and interpretation. Missing and incomplete data can be partially or completely compensated by using qualitative data expressed directly as fuzzy numbers.

Measured values (including those for socioeconomic components) are converted from crisp to fuzzy numbers in the environmental impact assessment model. Therefore, there is no reason to exclude other fuzzy numbers collected as qualitative observations that augment measured data. For example, drive-by surveys[2] can yield data such as "very many Canada geese were feeding on the pastureland," "there was an unpleasant odor detected along the western boundary of the site," "after last week's rains there are extensive erosion rills visible on the bare slopes," or "the noise from the nearby chemical plant was very loud at the base of the tree on which the bald eagles had built their nest." Without the use of fuzzy sets to capture potentially meaningful data on odors, noise, and relative abundances, these components would not be incorporated into the environmental impact assessment, and decisions based on the assessment will not be as well informed.

The use of such qualitative data was a valuable contribution to an assessment of changes to the seabed beneath pens of a salmon farm in the Red Sea ([32]). The seabed conditions under fish "corrals" is best described by the abundance and types of fauna and flora plus chemical analyses of sediment cores. Ideally, all collected data would be quantitative and carefully measured. But, with many pens, limited time, or rough seas, compromise was required. It was much quicker for a diver to record that the seabed had "very little seaweed, a few crabs, and thick patchy bacterial mats." To actually quantify that in terms of the biomass of seaweed, number of crabs per square meter, and thickness and percentage cover of mats requires special equipment and additional dive time; qualitative data—properly handled—is equally meaningful and valuable in assessing the impacts of the activity.

---

[2] Anecdotal reports from residents may also be used if deemed credible or of sufficient value. As will be seen in later sections such data will not unduly distort the results of the model.

The lead agency, or the professional preparing the assessment, sets up fuzzy term sets for each baseline component to be analyzed for change by the project alternatives. The included components are those that were identified during the scoping process as having the greatest importance to the majority of stakeholders. The baseline components are considered to be variables, and the term sets divide the Universe of Discourse into reasonable partitions. To do this so the resulting fuzzy sets represent the meaning of the linguistic variables they describe requires careful planning.

Developing the fuzzy sets (that is, determining the shape of the membership functions) lies at the intersection of science and values. On the one hand, assignment of an observed value to a grade of membership considered "adequate" represents societal values or the beliefs of stakeholders. On the other hand, subject matter experts are better able to interpret the adequacy of a measurement. The importance of distinguishing between stakeholders and experts is explained by [6]. They distinguish the roles of each by noting the similarities of stakeholders and experts in natural resource situations with those in legal trials. The stakeholder (or witness) confirms observations or events, but it is left to the expert to explain the meaning and significance of those observations or events. How this is applied in the development of fuzzy sets within an environmental impact assessment is dependent on the specific situation and location.

10.2 Fuzzy Set Design

Fuzzy sets represent knowledge on both sides of the IF-THEN rules in a fuzzy expert system used to conduct environmental impact assessments. Antecedent and consequent fuzzy sets share four components: Universe of Discourse, name convention, membership function, and thresholds [3]. Antecedent fuzzy sets also require linguistic hedges, linguistic input libraries, and fuzzifiers to convert the crisp input to grades of membership. Consequent fuzzy sets require normalization and defuzzification. Two skill areas are required of the human model designer: knowledge engineering and domain expertise.[3] For a com-

_____

[3] A knowledge engineer extracts from the domain expert the details needed to define both fuzzy sets and IF-THEN rules. Most often the knowledge en-

prehensive discussion of different methods of knowledge acquisition see [9].

This section introduces the components and considerations required to design fuzzy sets that are appropriately related to data variables, cover the appropriate range of input data, and reflect the semantics of the underlying concepts expressed by the linguistic variables.

Environmental conditions are modeled by fuzzy propositions— rules—that express the meaning of the observed components. Each component is a linguistic variable of concern or interest that is represented by a fuzzy term set. The term set must be decomposed into individual fuzzy sets, each of which reflects our understanding of that portion of the variable's range. The knowledge engineer determines the variables, term sets, characteristics of each fuzzy set, and the rules by interviewing the domain experts in detail. In the context of an environmental impact assessment model, the domain experts may include regulatory and resource agency staff, academic and consulting scientists (life and social), and residents of the area. (It will be shown during the discussion of rule generation that the interpretations or beliefs of so-called "dueling experts" can be accommodated within the modeling system.)

The Universe of Discourse defines the range of input values within which the fuzzy inference engine must be valid. Most input variables to a model are measures of length, area, time, and magnitude. The size of the project site, its distance from important landmarks, the number of weeks flowers are in bloom, large mammals or birds of prey give birth or fledge young, the number of jobs lost or created by the project and the change in daily traffic volume all may be input variables to an environmental impact assessment model. They all have different scales, meaningful numeric ranges, and the number of useful subcategories (fuzzy sets) into which they could be divided. For each variable, the measurement units and Universe of Discourse must be defined first.

The convention adopted for naming variables, included fuzzy sets and hedges, needs to be carefully considered. Well-thought-out names increase model comprehension, maintenance, and validation. The name

---

gineer works with a group of domain experts to develop the model that best represents the conditions, issues, and concerns of each specific environmental impact assessment.

of the linguistic variable should describe *what it is,* while the name of the fuzzy sets should describe *what it means.* It is helpful to think of fuzzy set names as the equivalent of language nouns and fuzzy set hedges as adverbs. Another consideration in naming fuzzy sets and hedges is the type of rule in which they are likely to be used. Rules can be *classifying* ("IF River_substrate is Bedrock THEN Headcutting is Negligible") or *prescribing* ("IF Slope is Steep THEN Erosion_risk is somewhat Increased"). In terms of an example used earlier, an input variable could be *Slope* and be defined by the term set *negligible, shallow, moderate, steep,* and *severe.* Appropriate hedges for this variable might include *somewhat* and *very;* the negation, *not,* would also be appropriate. The antecedents for a rule involving slope could have these structures—

```
IF Slope is not very steep ...
IF Slope is somewhat shallow ...
```

By convention, term sets have three to seven fuzzy sets into which they can meaningfully be divided. The number of fuzzy sets should also reflect both the quality of the underlying input data and their meaning in the assessment model. To illustrate, if the population size of a fish species of interest is determined by creel surveys, carcass counts, dam passage counts, or other methods that do not know the total population size then fuzzy sets of *small, medium,* and *large* population numbers would be more appropriate than partitioning the Universe of Discourse into more segments. The less precise sets better model the uncertainty in population size.

The most challenging part of designing fuzzy sets (and the entire term set for a variable) involves their shape, height, and position. While there is some experimental evidence [3] that the shape of the membership function may not matter, it is still important to devote sufficient effort to the shape so it closely reflects the underlying semantics. Within the broad context of environmental impact assessments, Gaussian curves better represent linguistic variables and fuzzy sets than do triangles or trapezoids. The latter two have sharp thresholds of change and constant slopes; most components of the natural world and associated human activities are not so constant. The interpretation of meaning for a triangular fuzzy set is that there is only one value of the input measurement in which the grade of membership is [1.0] and that greater and lesser values decrease in their grade of membership at a constant rate until [0.0]. Most variables of the natural, social, and economic world are much less precise. Their representation is better as a Gaussian curve, where there is a small range of values where the grade of membership is indistinguishably close to [1.0]; membership of values on either side decrease first slowly, then more rapidly and consistently (around the point of inflection of the curve), then more slowly again until [0.0]. Even values at the ends of the Universe of Discourse are more appropriately modeled by a membership function that gradually approaches or leaves the extreme values. For antecedent (input) fuzzy sets, each should have the maximum grade of membership ($\mu = [1.0]$) at its midpoint along the x-axis. Each membership function should have a support set width so the Universe of Discourse is equally partitioned into the number of fuzzy sets applied to the variable.

The other three considerations in membership function (fuzzy set) curve design are width, overlap and threshold value. Without specific reasons to justify unequal support set widths for each fuzzy set in a term set they should equally partition the Universe of Discourse. Each fuzzy set must also overlap its neighbor(s) so as to produce a continuous, smooth membership surface. There is no defined algorithm or procedure to establish the appropriate amount of overlap. However, experimental psychology studies in the early to mid-1800s by Weber and Fechner established a consistent point of *just noticeable difference* (jnd) [9]. When these two psychologists tested subjects to determine when changes in pitch, loudness, or color could be detected it was always about the 50th percentile. With equal heights and widths for each fuzzy set in the term, the [0.0] grade of membership of one set is at the same x-axis value as the [1.0] grade of membership of adjacent fuzzy sets.

Every linguistic variable's term set has a grade, or threshold level, below which any assignment to a fuzzy set is functionally meaningless. This α-level was explained above and is meaningful more in terms of consequent fuzzy sets than in antecedent fuzzy sets. Consider, for example, a solution set for the linguistic variable *Significant*. Within the context of the project, location, and socio-econo-environmental conditions it has been agreed that the term set consists of three fuzzy sets: *Slight*, *Moderate*, and *High*. Each fired rule contributes to the solution set, *Significant*, but if the output of the fired rule is below the threshold (α-cut), then it does not contribute to the solution. Details of this are shown in the worked example in Part III.

This explanation of creating linguistic variables and fuzzy sets explains how to augment sparse quantitative data with qualitative observations. When all observations are converted to grades of membership in a fuzzy set that describe a linguistic variable, the qualitative have the same computational weight as the quantitative.

10.3 Characterization

Section 6.3 on page 45 explains the deficiencies in the traditional approach to describing and characterizing the baseline conditions for an environmental impact assessment. The three deficiencies are:

1. Missing or inadequate data because of practical limits on detailed, quantitative data collection efforts.
2. Characterizing baseline conditions in a format comparable to the alternatives being considered.
3. Classifying baseline conditions on the subjective societal scale of acceptability.

These are deficiencies because they limit insight into the ecological dynamics of the existing conditions and limit the completeness of the environmental impact assessment.

All natural systems are highly dynamic. They undergo changes with periods from daily through seasonal and annual to decades or centuries. They also are subject to atypical changes caused by periods of drought or exceptional precipitation, earthquakes, fires, mudslides, hurricanes, volcanoes and other weather- and geologic-related phenomena. Incorporating quantitative and qualitative data on these events allows the assessment preparers to place the current landscape into perspective and to better anticipate both alternative effects and the changes expected under the "No Action" alternative. The susceptibility of the project site to infrequent natural impacts is a valuable project planning tool.

Characterization of baseline conditions reveals comparative quality, so both existing and future environmental conditions can be classified on the same scale. It is not valid to assume that the baseline conditions represent good quality in terms of wildlife habitats, wetlands, noise, or other environmental or socioeconomic components. The baseline conditions may well be rangeland that was overgrazed for decades, agricultural land with cattle grazing on the wetland vege-tation, or an urban zone or be subject to conditions generally thought of as undesirable. On the socioeconomic side, the area may be lack-ing in jobs or be in a developing region with untapped resources but poor infrastructure and less-than-desired administrative skills. When the baseline conditions are defined in a quantitative way, the site is placed on a quality scale relative to other alternatives and other base-line conditions.

Environmental impact assessments are subject to societal values of acceptability. Acceptability, of course, is a highly subjective term not directly measurable. But acceptability is the basis on which en-vironmental impact assessment decisions are made. Project alterna-tives are evaluated on their acceptability in terms of the significance of predicted change to the baseline conditions. Having the baseline conditions characterized by the same techniques as the alternatives, and ranked among them, puts the existing conditions in perspective. Because "acceptability" varies with geopolitical location, time, and project type, a discussion of approaches and methods is put off until consideration of decision-making (Chapter 12 on page 167). Because ecosystems are all different and the social, economic, and regulatory environments vary with locality and time, this section is not a cook book with explicit recipes to follow step-by-step. It presents a set of tools that can be applied to specific situations. There may also be situ-ations in which none of these tools are appropriate and new ones need to be forged.

The entry to existing condition characterization is the grade of membership in the appropriate fuzzy set of a linguistic variable.

Many environmental components have acceptability thresholds de-fined by statute or regulation. Water and air quality standards are in this category. At one level it is possible to declare that all standards must be met for the conditions to be considered acceptable. In fuzzy logic terms this is equivalent to combining grades of membership us-ing the AND operator (so that all must hold true for the entire state-ment to be true). The equivalent set operation is the intersection which is calculated as the minimum grade of membership. If water quality components are each quantified by the appropriate technique there are grades of membership representing how close the determined value is to the threshold of the standard. Figure 10.1 on the facing page shows that a measured 5 milligrams per liter of dissolved oxygen has a mem-

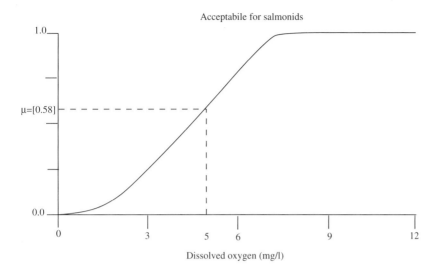

Fig. 10.1. Acceptable values of dissolved oxygen for salmonids, expressed as a fuzzy set.

bership in the fuzzy set, "acceptable for salmonids" of $\mu = [0.58]$. While this has limited value in characterizing the existing conditions when presented by itself, it gains in value when combined with other relevant factors. These factors might include rearing habitat quantity, macroinvertebrate food resource abundance, numbers of competing and predatory species and quantity of refugia. Because all these factors are necessary for "acceptable salmonid habitat" the overall value is calculated as $\mu_{fish} = min(\mu_{DO}, \mu_{habitat}, ..., \mu_n)$. In other words, the minimum value for all salmonid habitat components is the one used in characterizing the environment.

Many environmental conditions have no established regulatory standards. Neither do socioeconomic conditions. With these variables a given situation can be either acceptable or unacceptable depending on the perspective of the person or group evaluating them. As examples, in the Pacific Northwest of the United States logging was extremely restricted in the early 1990s because of reported numbers of spotted owls and marbled murrelets. From the conservationist ("environmentalist") point of view, this was a highly positive political decision. From the logger's (and logging communities') point of view, this was a devastatingly negative political decision. When there is no definitive "good" or "bad" to a variable the measures used to quantify them must work on either the or its complement; i.e., on A or $\neg A$. A highly useful operator to apply to these values is that of symmetric summation (Section 9.10.4 on page 104). Because it does not matter whether grades of membership come from a fuzzy set or its complement, components with different measurement scales and varying importance weights can be incorporated into the solution. The general equation for weighted symmetric summation of fuzzy sets is

$$\mu_{sum} = \left\{ [\mu_1/(1-\mu_1)]^A + [\mu_2/(1-\mu_2)]^B + \right.$$

$$\cdots$$

$$\left. + [mu_n/(1-\mu_n)] \right\}^{\frac{1}{A+B+\ldots+n}} \quad (10.1)$$

where μ_{xum} is the combined membership of all observed fuzzy sets (or their complements) and the relative weights assigned to each fuzzy set [32].

Another advantage of symmetric summation is that it accommodates missing data. Collecting data to establish baseline conditions or to monitor change over time involves multiple measurements during the series of collections. If equipment malfunction, adverse weather or other unusual event prevents the collection of a data value on occasion, the exponent of symmetric summation equation is adjusted to that of the number of included terms. These values can then be used in a time series analysis with value equal to every other data set. If equation ?? on page ?? represents data being collected for establishment of baseline conditions, and one sampling event could not collect component B, the partial data set can be represented by—

$$\mu_{sum} = \left\{ [\mu_1/(1-\mu_1)]^A + [\mu_3/(1-\mu_3)]^C \right\}^{\frac{1}{A+C}} \quad (10.2)$$

While this partial data set is not as informative as is a complete set, it is much better than no data for that collection date.

Over time, what initially is interpreted as a negative impact may turn out to have desirable benefits. An observation taught in most general ecology courses is that forest edges have more plant and animal species associated with them than do either the forest interior or the open meadow. A closed-canopy forest has all nutrients bound in living biomass and traps virtually all incoming solar energy in the canopy. This leaves the forest floor devoid of browse and cover plants for mammals. When an opening is created by nature or man, light and nutrients become available on the ground; and in a short time (tropics) or relatively short time (temperate zones) grasses, forbs, herbs, and other ground cover becomes abundant. Now mammals can find food, refuge from predation, and breeding sites. Whether the initial change in the forest composition was negative or positive depends on many factors, including one's perspective, the time frame and the context.

Silvert [32] raises an important consideration when combining different environmental components[4]: *complementary* versus *independent* fuzzy sets. Complementary fuzzy sets represent the range or extremes of a condition; a linguistic variable's term set consists of complementary fuzzy sets. Independent fuzzy sets describe two uncorrelated "adverbs" of the same "noun"; e.g., resident fish and predatory fish. One species of fish may be both resident and predatory, but other species may be one or the other. In the context of an environmental impact assessment, both complementary and independent fuzzy sets compose the description of environmental conditions. The amount of wildlife habitat might be described as "somewhat moderately high" because the grade of membership associated with the areal size is associated with the two fuzzy sets, "moderately_high" and "high," but with a higher grade in the former term. When the quantities of "traffic_volume" and "noise" are considered, they are independent fuzzy sets that both affect the acceptability of the environment.

There is a lot of flexibility in using all the available tools to characterize the existing conditions (and the projected alternative conditions) to accommodate different projects, locales, and political preferences. The recommended approach is to create appropriate fuzzy sets for each of the components identified as important in the scoping process and calculate the weighted geometric mean of the grade of membership for a component raised to the power of its relative weight—

---

[4] This concept is also applicable to evaluating crisp measures of environmental components but is only rarely incorporated into the analyses.

$$\bar{G} = \prod_{i=1}^{n}(\mu_i^{w_i})\tag{10.3}$$

The geometric mean decreases the influence of the extreme values (high and low) while increasing with greater membership grades for the components. It also grants greater influence to those components weighted most heavily by consensus of all stakeholders, experts and other participants in the assessment process. The grades of membership (μ) are each in the set [0,1] and the proportionate weights must add up to 1.0.

The process of creating appropriate fuzzy sets to characterize environments follows the same path regardless of the component or measurement scale; this process was described in general terms in the preceding section. Specific applications are documented in Part III on page 179.

With components having spatial attributes of value (e.g., wildlife habitats, wetlands) the fuzzy term sets might be labeled with reference to areal size as *Very_small*, *Small*, *Moderate*, *Large*, and *Very_large*. The universe of domain could range from 0 hectares to 5 kilometers square scaled, of course, to the size of the project and the anticipated impact area. The specific measured values (or best professional judgments) of size are fuzzified into a numeric grade of membership. Wildlife habitat or wetland quality could be defined as separate linguistic variables as *Low*, *Moderate*, and *High* with each term having a range based on expert judgment of defined criteria or an arbitrary scale acceptable to the majority of decision-makers.

Purely conceptual linguistic variables are always part of an environmental impact assessment. For these variables, quantification is made easier by following a two-step process to create the appropriate Type-2 fuzzy set. Consider the linguistic variables of *Scenic_value* and *Recreational_opportunities*. These come up frequently in assessments under these or similar terms (e.g., "open spaces"). While each of these linguistic variables can be decomposed into components, and each component measured, fuzzified, and used in a rule-base processor to create the output fuzzy set, the results are not worth the effort. These pure linguistic variables reflect values and beliefs and cannot be made more "accurate" by extended manipulations. The survey approach to the first step works very well [20]. The history of this technique supports its use in describing all antecedent pure linguistic variables in environmental impact assessments.

Mendel and a social scientist colleague (experienced with survey methods) established a list of 16 words and phrases that most people would use to describe a set of 0—10 objects. A group of engineering students was presented with the random-ordered list and asked to put beginning and ending values on each term within the stated range. From the 70 valid surveys returned (some were disqualified because the entire range was entered for each label), they calculated the mean and standard deviation for the starting and ending values of each label. Continuing their analyses to determine the minimum number of labels needed to cover the range 0—10, they conducted a second survey (n = 47 respondents) and determined that five labels (with the standard deviations) provided adequate overlap. Table 10.1 reproduces Table 2-3 in [20]. What these results confirm is that words really do mean different

Table 10.1. Word variable labels and parameters for Type-2 fuzzy sets.

		Mean		Std. Dev.	
No.	Range Label	Start	End	Start	End
1	*None to very little* (NVL)	0	1.9850	0	0.8104
2	*Some* (S)	2.5433	5.2500	0.9066	1.3693
3	*A moderate amount* (MOA)	3.6433	6.4567	0.8842	0.8557
4	*A large amount* (LA)	6.4833	8.7500	0.7484	0.5981
5	*A maximum amount* (MAA)	8.5500	10	0.7468	0

things to different people.

To create the fuzzy grade of membership for the environmental component, *Aesthetics*, for example, build the term set for the linguistic variable using the values of Table 10.1. The mean values for the start and end represent the middle of the fuzzy set, while the standard deviations around these values define the FOU (Footprint of Uncertainty). Figure 10.2 shows this for the middle fuzzy set, *a moderate amount*. The other term sets would overlap this one by 50 percent. Notice that the ends have no FOU extending less than 0 or greater than 10. Once the term set is defined by stakeholders, decision-makers and the public can be polled for their votes on the æsthetics of the existing condition. By

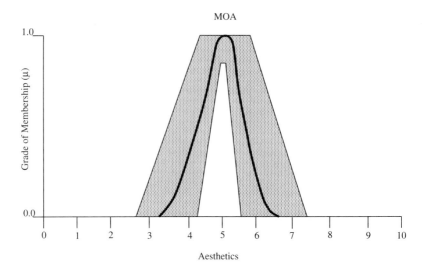

Fig. 10.2. The Type-2 fuzzy set for *a moderate amount*. This is the middle set of five in the term set for the linguistic variable, *Aesthetics*.

asking for opinions based on the word labels, we can apportion the results among those labels and the average used as the entry point (on the x-axis). The grade of membership is then the midpoint of the intersection of a vertical line from that point on the x-axis to the lower and upper bounds of the FOU.

The same process—using the same word labels—can be used to establish the acceptability of the existing conditions and that of the various project alternatives. Alternatively, the concept of acceptability can be built from the evaluation of rules that evaluate the grades of membership of assessment components that contribute to acceptability. These components will include physical and biological considerations, socioeconomic conditions, and the broad landscape in which the project is located. There is a lot of flexibility in adapting these tools to specific projects, specific locations, and the prevalent societal values.

11

Impact Inference and Assessment

11.1 Societal Values and Beliefs

Much effort in the assessment of actual or potential environmental impacts is devoted to matching science with societal values and beliefs. Countries, regions, and cultures have different values or beliefs. While everyone agrees there are concepts of significance, acceptability, and sustainability there may not be agreement on how to define these terms. In Chapter 6.3 on page 45, the various attempts to apply the concept of significance to environmental impact assessments were summarized. In this chapter significance, acceptability, and sustainability are considered in the context of imprecise variables that can be quantified and manipulated objectively and rigorously.

The choice of environmental components to be included in an impact assessment is a combination of statutory requirements, the professional judgments of the technical team preparing the assessment on behalf of the project developer and resource agencies, and a reflection of general societal values. It is in the scoping process that the public and other stakeholders get to express their values. The choices made by a team of technical experts may well differ from the choices made by lawmakers, regulators, or other stakeholders. As early as the scoping stage of an assessment different values can set a confrontational tone that remains for the duration of the assessment. In the explanation of the scoping process (Chapter 3 on page 23) it is noted there are a number of advantages to the project developer, regulators, impact assessors, and decision-makers when public comments are solicited at an early stage of the assessment process. Those who are supportive or neutral to the project can bring their local familiarity and different viewpoints to creating the list of environmental components and alternatives to be considered. Those who object to the project will have an opportunity to express their opposition and have it incorporated into the analysis. In other words, different societal segments may express their values and beliefs regarding the project.

The objective is to produce a list of environmental components that are ordered in importance and include the beliefs and value systems of everyone. The solution uses the ideas of Saaty [28] for calculating a ratio scale of pairwise comparisons for the set of elements under consideration. As extended by Yager [41], this calculation produces a vector of exponential weights reflecting relative values of the components.

The method involves each interest group with common values and beliefs choosing one from each pair of components—the one that has the higher value *to them*. The group then determines *by how much* the one component is of higher value. These values were carefully chosen to be meaningful, comprehensive, and easy to use (Table 11.1). These

Table 11.1. Importance values to be used in pairwise comparisons of environmental components.

Importance value	Definition
1	Equal importance
3	Weak importance of one over the other
5	Strong importance of one over the other
7	Demonstrated importance of one over the other
9	Absolute importance of one over the other
2, 4, 6, 8	Intermediate values between the two adjacent definitions

values are placed in a square, symmetrical matrix, with each cell representing the collective values of the group for a specified comparison. For example, if wildlife habitat (i) is more important than water quality (j), then the importance weights are entered in the matrix as $a_{ij} = 1/a_{ji}$; the reciprocity is necessary to maintain the square matrix.

The maximum eigenvector is multiplied by the matrix order (*i.e.*, the number of rows/columns in the square matrix) to create weights of the fuzzy sets being compared [24]. The result is a vector whose elements reflect the collective relative value of each environmental component. With this rank-ordered list the lead agency can select those components to be evaluated in the assessment from among all components. Whether this decision is made at a break in the sequence of values or limited to the top N components, the selection is objective rather than arbitrary or capricious. Also, everyone's opinions were considered, and considered equally. There was no special processing or treatment that gave one set of values greater weight than another set of values.

To illustrate this procedure, consider the greatly simplified example of three environmental components: wildlife habitat (H), wetlands (W), and water quality (Q). One group of stakeholders believes that—

- Wetlands (W) are weakly more important than wildlife habitat (H); so $a_{HW} = \frac{1}{3}$ and $a_{WH} = 3$.
- Water quality (Q) is somewhere between equal and weakly more important than wildlife habitat (H); so $a_{HQ} = \frac{1}{2}$ and $a_{QH} = 2$.
- Wetlands (W) are weakly more important than water quality (Q); so $a_{WQ} = 3$ and $a_{QW} = \frac{1}{3}$.

The resulting matrix of the paired comparisons, M, is

$$M = \begin{bmatrix} 1 & \frac{1}{3} & \frac{1}{2} \\ 3 & 1 & 3 \\ 2 & \frac{1}{3} & 1 \end{bmatrix} \cdot \begin{bmatrix} H \\ W \\ Q \end{bmatrix} \tag{11.1}$$

To calculate the eigenvalue and eigenvector for this matrix use the equation, $MW = \lambda_{max}W$ to obtain the normalized eigenvector (for λ_{max})

$$W = \begin{bmatrix} 0.16 \\ 0.59 \\ 0.25 \end{bmatrix} \tag{11.2}$$

The interpretation of this example is that the stakeholder group believes that wetlands (W) are more than twice as important as water quality (Q), and wildlife habitat(H) is the least important. These differences are based on very small belief differences as described in the paragraph immediately above the matrices.

This method is a solution to the problems of incorporating public input and determining relative importance values of the set of components included in the EIA. Requiring pairwise comparisons of all components promotes awareness that there is no way to have everything, even with the "No Action" alternative. There are always compromises that must be made in the real world, and this brings everyone into the decision-making process. The ranked environmental components also indicate where time, effort, and money needs to be apportioned to the data collection portion of the EIA process. In the example above, much less effort should be devoted to characterizing wildlife habitat than is devoted to water quality and wetlands.

Applying this procedure to the scoping process is a technically sound, legally defensible way of determining what issues, concerns, and components to include in the environmental impact assessment, while including everyone's vested interests. This ranking of values is used during characterization of the baseline environmental conditions, when describing alternatives and when evaluating alternatives.for a decision. For these uses, the eigenvector weights will be normalized by multiplying each weight value by the number of components (3, in the example above) and using the resulting values as exponents to calculate an ordered weighted average (OWA) or minimum entropy-ordered weighted average (ME-OWA) for multi-objective, multi-criteria decision-making. These OWA will be used to rank the future environments from all of the alternatives.

However, three of the most common and important societal values are significance, acceptability, and sustainability. Each of the three is a word-concept linguistic variable; that is, there is no way to measure them directly. Therefore, they cannot easily be used as antecedent fuzzy sets (i.e., on the left-hand-side of an IF-THEN rule), but they frequently are the consequent fuzzy sets generated by those rules. The major difference between significance/acceptability, and sustainability is that the former can be based on indirectly measurable values while the latter cannot.[1] The procedure for determining these three societal values when evaluating project impacts involves the development and use of a fuzzy expert system model that appropriately describes the project and location.

11.2 Fuzzy Expert System Models

Before, most decision criteria could be calculated with the appropriate mathematical approach. Many applications of fuzzy expert systems are constructed using only mathematical equations (e.g., operations research models on optimal production levels for factories; setting stockholder dividends based on several competing criteria). Mathematical models also are well suited to identifying the consensus set of environmental conditions during the scoping phase of an environmental impact assessment. However, all other phases in environmental impact assessments are based on the experience of experts in addition to societal values and beliefs. Models that succeed in the environmental impact assessment process replicate the reasoning process used by an expert (or multiple experts, even those with conflicting opinions) in the subject to infer conclusions from antecedent knowledge. This category of computational intelligence consists of approximate reasoning expert system models.

Approximate reasoning models have five main steps: input data processing, evaluating antecedent fuzzy variables, fuzzy inference engine processing, defuzzification, and solution presentation. (Steps are numbered 1—5 in Figure 11.1 on the next page). Steps 1, 2, 4, and 5 were introduced in Chapter 9 on page 63. However, when Type-2 fuzzy sets are used, defuzzification requires a two-step process to yield a crisp, numeric result. The first step is to simplify the Type-2 fuzzy set into an equivalent Type-1 fuzzy set. The latter is then defuzzified using the most suitable method (generally center of gravity/center of moments).

Within the context of developing the model for a specific environmental impact assessment, each of the steps is summarized. The "black

---

[1] This may well change in the future, but at the time this is being written there is no consensus definition of "sustainable." It has been described as a process or a goal with arbitrarily defined values. Sustainability may be the most complex societal value to quantify, regardless of method.

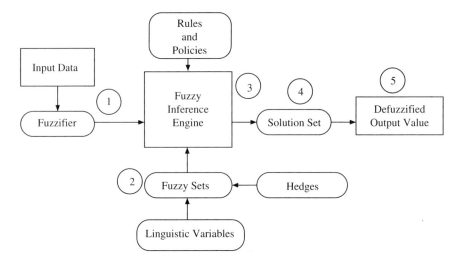

Fig. 11.1. The major components of an approximate reasoning expert system model and the flow of processing through the components.

box" labeled "Fuzzy Inference Engine" is the heart of the fuzzy expert system used to assess the significance, acceptability, and sustainability of different project alternatives. This step is more fully described because it was not previously introduced.

11.2.1 Designing the Model

From the earliest days of artificial intelligence (AI) in the 1970s and 1980s, expert systems have been most commonly available as "shells." An expert system shell is a program that facilitates the development of expert system applications. The concept is analogous to that of a database management system that facilitates the development of a database application. Regardless of the tool selected for use in a given situation, the environmental impact assessment fuzzy expert system model must be planned.

For each environmental impact assessment, linguistic variables, fuzzy term sets, hedges, and output needs are defined by location, project type, statutory and regulatory environment, socioeconomic setting, and environmental landscape. During this solution strategy development it is also necessary to identify those who have the requisite, pertinent knowledge that will be fed into the fuzzy inference engine in the form of IF-THEN rules. These rules are part of what defines significance, acceptable, and sustainable.

Fuzzy system models have a very high degree of freedom because they represent subjectivity and individual styles, beliefs, and values. As a result, the objective of these models are solutions that are "good enough" approximations to make informed decisions. And, because of the structure and inherent imprecision of the environmental impact assessment process, numeric models using an engineering approach cannot provide the same degree of confidence and comfort in the derived solutions. Regardless, the design process must be iterative because the components are interdependent. The fuzzy variables in the model are dependent on the fuzzy rules and project-specific decision boundaries. In turn, the rules may require modification of the linguistic variable term sets and membership function shapes.

11.2.1.1 Universe of Discourse

The numeric range of each variable (both antecedent and consequent) must be determined. For antecedent variables, the number of fuzzy sets in the term set are directly related to the Universe of Discourse. In some cases the individual components of the term set are determined first, and their range of values then defines the Universe of Discourse. Otherwise, the term set is fit within the Universe of Discourse. Antecedent and result variables have a stated acceptable range of values that is the Universe of Discourse, but there is no term set on the right-hand side of the IF-THEN statement.

11.2.1.2 Name Convention

Selecting names for variables and individual fuzzy sets in the term set requires careful consideration. Names must be descriptive, appropriate to the consequent and antecedent variables as well as the hedges that might be applied. The names of individual terms in variables do not need to be unique (that is, a number of variables can have term sets of "small," "medium," and "large") but variable names must be unique. Antecedent variables do not have separate hedges applied to them. However, a domain expert may evaluate the result of an action as "very much true." In these cases, the resulting variable will be named that way, and its shape and Universe of Discourse will incorporate the modifying terms.

11.2.1.3 Hedge

Linguistic hedges modify membership functions of fuzzy sets by increasing or decreasing the range of the support set. There are established computations for specific hedges, but the definitions can always be adapted to specific needs. It is generally a very good idea to develop a set of hedges to use in the environmental impact assessment model and use those for consistency with all variables.

11.2.1.4 Membership Function Shape, Overlap, and Location

There is some experimental evidence that suggests most real-world fuzzy system models are relatively insensitive to the shape of the membership function. However, there is sufficient difference in semantic meaning of terms in different countries or regions that careful consideration needs to be given to how each fuzzy set is defined. For most variables in an environmental impact assessment the Gaussian curve is a good starting point. Experience and sensitivity analysis of the model will allow the shape to be tuned if necessary. Similarly, overlap among adjacent fuzzy term sets in a variable is generally set at 50 percent; almost always the overlap is in the range of 35–65 percent. The number of fuzzy term sets in a linguistic variable will determine their relative locations and support set range. Most linguistic variables have three to seven fuzzy term sets. The number is dependent on the granularity, or resolution, deemed appropriate for that linguistic variable in the environmental impact assessment model being applied.

Consequent membership function shape must also be determined in advance, but there are no fuzzy term sets within it. Modifications to the shape and height of the solution fuzzy set are made by the parallel processing of antecedent rule implication and aggregation. The lower threshold for meaningful height of the solution fuzzy set can be set by a defined α-level or by an unconditional rule.

11.2.1.5 Defuzzification and Output Processing

The defuzzification method of center of gravity will be appropriate for almost all environmental impact assessments. However, the translation of a resulting crisp value into a linguistic form must be part of the model design process. It is at this stage that values are assigned to concepts such as significance and acceptability.

11.2.2 Rule Creation

Neither this book nor any other book can present a fixed set of rules applicable to all environmental impact assessments, nor to any specific one. However, there are guidelines that make the rule development process easier and more robust. Two important guidelines discussed are input data sources (experts and data collections) and organizing rules in a large, complex model. Other considerations include the logic, implication and aggregation operators to be used and importance weights.

11.2.2.1 Sources

11.2.2.1.1 *Knowledge Engineering*

Two categories of "experts" are most commonly consulted when developing rules and relations in an environmental impact assessment fuzzy system model. For most of the environmental categories the subject matter (domain) expert; biologists, ecologists, hydrologists, engineers, and economists are common examples. Other sources of information include published articles, agency reports and university theses or dissertations. The purpose of knowledge engineering is to extract the actual decision-making process given certain input conditions. The procedure used by the knowledge engineer is based on both narrative and interview (that is, questions and answers).

The narrative involves the subject matter expert's describing how he or she assigns values to observed or measured environmental conditions. The interview extends (or replaces) the narrative by having the knowledge engineer elicit details and clarifications on the entire decision-making process of the subject matter expert. There are books, journals, articles, and world wide web pages devoted to knowledge engineering. Good introductions to the subject are also provided by [3, 9].

11.2.2.1.2 Data Collections

Many of the environmental components in an environmental impact assessment consist of measurements that may have both temporal and spatial features associated with them. While this book does not cover statistical methods, such analyses are very useful in defining rules as well as Universes of Discourse and attributes of fuzzy term sets.

Precipitation data are often available for a long period at or near the assessment site. In addition to extracting statistics such as mean, range and standard deviation from these data correlations are often possible. For example, records may indicate that some animal populations increase in years of high precipitation because there is greater plant food or cover available. Other animal populations may not fluctuate with precipitation patterns. Depending on the area involved in the assessment, its proximity to similar areas, and the distribution of animal species populations of interest, the project may or may not have a direct impact on the species. Consider a well-vegetated draw in the north-facing slope of a semi-arid region that is within the assessment area. If the quantity of vegetation (and its nutritive content) is a reflection of the amount of winter precipitation and rate of snow melt, and it is the only such patch for a relatively large distance (notice the fuzzy terms used to describe this situation), then animal populations may be greatly decreased if the vegetation is removed.

Similarly, in a headwater tributary of a river system the success of salmonid fish spawning (as measured by the percentage of eggs that produce live alevins) is dependent (in part) on the dissolved oxygen levels of interstitial water flowing through the gravel streambed. A rule could be developed from these data:

```
IF dissolved_oxygen is slightly high,
   THEN egg_hatch is reduced.
```

11.2.2.1.3 Voted-For Distributions

This technique can be applied to any variable in an environmental impact assessment but it is particularly well suited for the purely linguistic (Type-2) variables. There are no objective criteria that can be applied to concepts such as *Significant, Acceptable* or *Sustainable*. However, a population of expert voters—or groups of stakeholders—can classify each value or range of values within the Universe of Discourse. A group of wildlife biologists, for example, can be asked how many animals in a certain species they would consider to be *Significant* as a population at a specific location. For a set of numbers that might encompass the possible population size at that site, the biologists are asked to indicate whether they consider certain numbers as *Significant*. The percentage of biologists who vote for each offered number represents the grade of membership of that number in the fuzzy set *Significant* (Figure 11.2a). Connecting the point values with a line produces the membership function based on the values assigned by the biologists (Figure 11.2b).

The voting-for technique can be extended by applying weights to the value of each expert's opinion. It can also be extended by using it with nonexperts, too. For example, residents in the vicinity of the proposed project could be asked to vote on the acceptability of different commuting times, traffic volumes, noise levels, and other components of the environmental impact assessment. Even with variables that do not have a true measurement base, people can vote on relative values along an arbitrary scale of, for example, 0–100. This encourages people to think about their values and how they would assess changes in *Significance*, *Acceptability,* and similar concepts along a ratio scale. This is not much different from asking people to assign a grade of membership in the fuzzy set *Uncomfortable* based on the ambient air temperature.

11.2.2.2 Rule Organization

Environmental impact assessments involve many components spanning the spectra of environmental, socioeconomic, physical, and values considerations. No matter what approach is used by the practitioner, there is a very large volume of data to organize and integrate. When a fuzzy system model is built to take advantage of computational intelligence in the assessment process there can be hundreds of rules defining relationships. Without organization of the rule base, the model will not produce useful results because rules are fired in parallel to produce a solution set. If all rules in the model were applied simultaneously, the output would be uninterpretable. The solution is to partition the rules into logical groups called *policies* [9].

The complexity of an environmental impact assessment lends itself to a policy hierarchy. Top-level policies could be Environment, So-

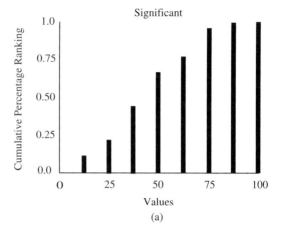

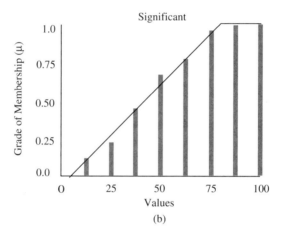

Fig. 11.2. Voting frequency by wildlife biologists on whether specified population sizes are significant.

cial, and Economic. Subpolicies for Social would include transportation, recreational opportunities, visual æsthetics, housing, schools, and open spaces. Economic subpolicies would include jobs, tax base, public services, and population growth. Within the environment policy would be a sublevel of terrestrial, aquatic, and wetlands. At the next level could be policies for animals, plants, air (or water) quality, land use, geology, topography, and precipitation. Additional policy levels can be created as needed to address all concerns of the project adequately. Other rule formation strategies include competitive and cooperative, but the hierarchical strategy is best suited to the objectives of environmental impact assessments.

The granularity of the policies should track the granularity of issues considered during the scoping stage of the environmental impact assessment process. If, during scoping, all assessment components are ranked for importance by all stakeholders and other interest groups, it will be relatively simple to assign both significance and priorities to the output of each policy.

Fuzzy system models consist of both unconditional and conditional rules. If a policy contains only unconditional rules, or only conditional rules, then the evaluation order is immaterial. However, environmental impact assessment models are in the class of fuzzy system models that contain both types of fuzzy propositions within a single policy. In these cases, the sequence in which the rules are evaluated (or "fired" in fuzzy system model parlance) matters because the nature of the solution fuzzy set depends on whether unconditional rules are evaluated before or after the conditional rules.

The unconditional rules are used to establish the defaults for the solution fuzzy set; that is, they generally define the minimum or maximum values that the solution can attain. This means that there is a solution even if none of the conditional rules in the policy is fired. Therefore, environmental impact assessment models will evaluate unconditional rules and apply them to the solution fuzzy set before any of the conditional rules are evaluated. Also, if no fired conditional rule has a strength greater than that of the strongest unconditional rule, then the conditionals do not contribute to the output. This also means that if a fired conditional rule is stronger than any unconditional rule the latter do not contribute to the solution fuzzy set. There is no way to predict the outcome; the rules in the policy must be evaluated with the supplied input data to determine the outcome.

An example would be from an environmental impact assessment of a proposed dredging program for flood control purposes in the city of Tillamook, Oregon. Three rivers (the Trask, Wilson and Kilchis) flow into the Pacific Ocean within the city limits. Sediment deposition in the river deltas causes extensive flooding during winter rains. Coastal Coho salmon (*Oncorhynchus kisutch*) and other salmon populations are listed under the Endangered Species Act, and some stakeholders believe that dredging of the original channels will indirectly harm fish by altering habitats, behavior generally prohibited under the ESA. Those expressing opinions to be considered in the assessment include city officials, environmental groups, federal and state resource and regulatory agencies, the tourist industry, residents, and local businesses. These stakeholder groups have different levels of vested interests in various project alternatives.

One scenario for a rule policy addressing salmonid habitats might include these rules:

1. The habitat values of these estuaries must be low.
2. The habitat values of these estuaries must be high.
3. If dredging historic channels allows better fish access to the upper river channels, then estuary habitat values can be somewhat low.
4. If fish populations in the estuaries have increased since they filled with sediments, then estuary habitat values are high.
5. If fish populations show no change in size variation over the past four decades, then estuary habitat values are moderate.

The first two rules are unconditional; the last three are conditional. The unconditional rules establish minimum and maximum values for the consequent fuzzy set, "Estuary_habitat_values" (see Figure 9.22 on page 103 to follow how different antecedent values contribute to a resulting fuzzy set). When the conditional rules are evaluated, they have no contribution to the result if they are greater than the unconditional maximum value or less than the unconditional minimum value. They will, however, alter the shape of the consequent fuzzy set, so when it is defuzzified to obtain the crisp value for the variable, each of the five rules has contributed to the result.

11.2.2.3 Logic Operators

Fuzzy logic provides both standard and nonstandard versions of operators such as AND, OR, and NOT. In the fuzzy logic literature, these operators are sometimes called aggregation, algebraic, compensatory, or composition operators. The function of logic operators is to determine the logical outcome of combining multiple membership (or truth) values. In Boolean (crisp) logic, the two basic operators are AND and OR, corresponding to the set operations of intersection and union, respectively. In fuzzy logic they also apply to intersection and union, but their mathematical expression is not fixed or limited. Fuzzy logic theoreticians have developed the concepts of triangular norm (*t-norm*) and triangular co-norm (*s-norm*), which encompass the multiple methods presented in the literature to calculate AND and OR. In almost all fuzzy system models for environmental impact assessment the most common operators are used.

The AND logic operator is equivalent to the fuzzy set intersection operation. It is the minimum membership value of the fuzzy sets being combined. The most common AND operator (the basic Zadeh AND operator) is the minimum: $min(\mu_A[x], \mu_B[y])$. There is an alternative operator with attributes useful in many situations: the product AND operator. Rather than the minimum degree of membership as the result, it is the product of the two degrees of membership that is used: $prod(\mu_A[x] * \mu_B[y])$. The attribute of importance of the product AND operator is that the importance of low values for both variables is greatly reduced. This result is useful when the min-max aggregation operator is applied. Suppressing the value of two small grades of membership ensures that nonimportant variables do not appear in the results of a computation. The product AND operator is one of the class of *compensatory operators*. Compensatory operators are those discussed in Section 9.7 on page 92.

The basic Zadeh OR operator is equivalent to the fuzzy set union operation. It is the maximum value of the degrees of membership of each fuzzy set being combined: $max(\mu_A[x], \mu_B[y])$. There is no product version of the OR operator, but there are compensatory operators which are along the continuum between the extremes of AND and OR.

11.2.2.4 Implication Operators

The two logical inference methods most useful in environmental impact assessment models (modus ponens and modus tollens) were defined and explained in Section 9.10 on page 99. The choice of which implication operator to use is very important, but there are few, if any, guidelines in the literature. Because environmental impact assessments are large and complex, the approximate reasoning models will use both operators in different stages of the assessment process.

There are three broad inferencing schemes applicable to rule formation in an approximate model: relational inference, proportional inference, and compositional inference. The approach chosen to form rules for a model do not depend on the type of problem to be solved (e.g., control, diagnostic, classification, prediction/forecasting) but on the nature of the problem. All three schemes are used in environmental impact assessments where the types of problems addressed vary widely.

11.2.2.4.1 Relational Inference

Relational inference directly associates elements of two or more fuzzy sets. That is, relational inference is governed by unconditional rules such as *Stream_spawning salmon ARE protected from harvest*, *The bald eagle nest IS near the parking lot*, or *Spotted_owls ARE very_similar_to barred_owls*. The wording of relational inference statements is based on a limited vocabulary which is repeated with different associated objects; the examples use statements of being (*IS, ARE*). Relational inference is also very commonly used in fuzzy system models that solve diagnostic problems such as why a car will not start or whether a patient might have a mental illness. In environmental impact assessment models fuzzy inference is usually limited to unconditional rules that express either extreme values (*No wetlands can be filled or removed*) or conflicting expert opinions (*The population size of wolves is self-sustaining* and *the population size of wolves is so small they are threatened with extirpation*).

11.2.2.4.2 Proportional (Monotonic) Inference

This inference method is used in limited, specific circumstances in fuzzy system models (including those for environmental impact assessments). The method is proportional because of the direct connection between input variable value and output variable value. Also, as the input value increases or decreases, the output value similarly increases or decreases. Therefore, the two variables always change in the same direction—that is, monotonically.

A very simple example is the relationship between assessment area size and assessment cost. They are related through the proportional implication function, *if a is L then c is H*. Mathematically this is expressed as

$$c = f((a, L), H) \qquad (11.3)$$

This is read that the cost is proportional to the area as expressed in the membership function shapes of the linguistic variables L and H.

11.2.2.4.3 Compositional Inference

Computational inference is the most generalized inference scheme that combines many antecedent rules, in parallel, to form the solution set. Most problems (including environmental impact assessments) are structured with relationships among the antecedent linguistic variables (and among their fuzzy term sets). In other words, "composition must connect individual relational inferences through a linguistic framework" [3]. The inference mechanism includes an implication method (such as min-max or fuzzy additive, discussed in Sections 9.10.1 on page 101 and 9.10.2 on page 102), aggregation and defuzzification. The compositional inference method is illustrated in Figure 9.22 on page 103.

The action of both the min-max and fuzzy additive implication methods is to reduce the truth of the consequent fuzzy set so that it is no greater than the truth of the rules. This is done before the solution variable's fuzzy set is updated by means of the aggregation operator.

11.2.2.5 Aggregation Operator

Aggregation operators are also called correlation methods. These are used in the process of combining the implication results to produce a meaningful and explainable solution fuzzy set. The action of the aggregation operator is to limit the height of the consequent fuzzy set so that its degree of membership (or truth value) reflects all the antecedent rules in that policy. There are very few aggregation operators available; the two most commonly applied in models of the type that includes environmental impact assessments are the correlation minimum and correlation product. The solution fuzzy set aggregated by the fuzzy inference engine is the final result of application of fuzzy logic theory. Additional processing to a scalar value, and interpretation of the results in the context of the environmental impact assessment specifics, require use of additional tools of computational intelligence and decision-making.

11.2.2.5.1 Correlation Minimum

The correlation minimum is the prevalent method of correlating consequent truth with antecedent truth. The method works by truncating the consequent fuzzy set at the maximum degree of membership of the antecedents minimum degrees of membership. In other words, the consequent fuzzy set's height (degree of membership, truth value) is *minimized* to the maximum truth value of the antecedent's in the rules. This correlation method is illustrated in Figure 9.22 on page 103. In this illustration, the truth value for erosion risk is minimized at [0.48], and this becomes the current value available to be used in further processing.

11.2.2.5.2 Correlation Product

There are situations where the correlation product method is better than the correlation minimum method at representing the meaning of the intermediate fuzzy set's truth. The truth of the consequent fuzzy set is represented by scaling it to the maximum value of the antecedent minimum degrees of membership rather than by truncating it. Scaling retains the shape of the consequent fuzzy set (Figure 11.3 on the facing page).

The advantage of the correlation products aggregation method is that it does not introduce plateaus on the consequent fuzzy set. However, the disadvantage is that irregularities in shape can be introduced and these may have effects on the defuzzified outcome when the composite moments or composite maximum methods are used. The trade-off in methods can also be expressed as reduction in information loss with the correlation products method versus potential change in defuzzified output value as the output variable's fuzzy set continues to be developed.

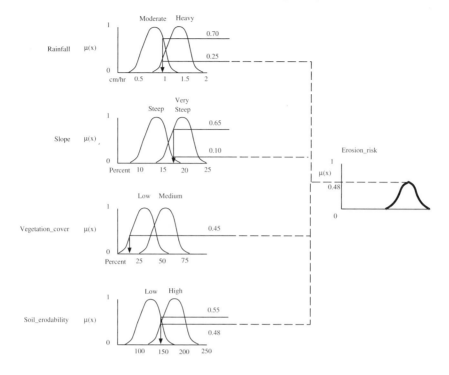

Fig. 11.3. Correlation product of erosion risk.

11.2.2.6 Defuzzification Method

Defuzzification, or decomposition of the solution fuzzy set, was introduced in Section 9.9 on page 98. It is the process by which implication results are converted from the range of possibilities represented by the solution fuzzy set to a scalar value in that Universe of Discourse. As explained earlier, defuzzification is inherently an approximation of the aggregated implication of the rules despite having a crisp value. Approximating large or complex sets of numbers with a single value is the mathematical basis for statistics and is not at all a limiting factor in the use of fuzzy sets and fuzzy logic in the assessment of environmental impacts. Berkan and Trubatch [3] explain the role of defuzzification very well: "When the aggregation process is viewed as the contribution of individual decisions, then defuzzification can be viewed as acquiring a popular vote or consensus."

The *center of gravity* (or *composite moments*) method is the most frequently applied defuzzification technique (equation in Section 9.9 on page 98). It represents the balance point of the fuzzy set and is equivalent to the statistical mean value of a population sample. The center of gravity method is usually applied after union aggregation of the rules in that policy block.

Another method used in many models is that of the *maximum possibility* (or *composite maximum*). The solution fuzzy set represents a possibility space so the value within the Universe of Discourse at the maximum height of that fuzzy set (i.e., the greatest degree of membership) is the most possible value. When the solution fuzzy set has a plateau then either the mid-point of the plateau or the average of maximums is used to determine the scalar maximum possibility value. Maximum possibility is also applied to union aggregation.

When the aggregation operator uses intersection the *center of mass* (or *preponderance of evidence*) decomposition method locates the region with the highest density of intersecting fuzzy sets. In practice, this defuzzification technique is performed simultaneously with the inference process, because it is necessary to track the location of each partial-solution fuzzy set. The result is determined by counting frequencies of included solution sets or by taking the maximum point of possibility of the intersected sets.

Other decomposition methods have been presented in the approximate reasoning literature. These include the *far and near edges of the support set*, the *center of maximums*, and *singleton geometric representations*. Care is needed when considering the use of any of these to ensure that they are appropriate to the underlying semantics of the solution fuzzy set. As noted above, the composite moments method is very well suited to environmental impact assessments and should be the defuzzification technique used unless compelling reasons justify a different one.

11.2.2.7 Importance Weights

The rule base for a fuzzy environmental impact assessment model can be large because there are many components to be evaluated. The rules for the components are organized into policies or blocks, but each policy has a different importance value. This was determined during the scoping process when participants explicitly ranked the set of components by consensus value. These differences in importance are reflected in relative weights assigned to each policy. In addition, the individual rules within each policy also vary in importance and are assigned weights to reflect these differences. Virtually all real-world fuzzy system models require different weights to more accurately reflect the underlying decision processes [44].

With an appropriate weight on each rule and policy it is possible to conduct sensitivity analyses of the model [3]. The results provide insight into both the dynamics of the model process itself and the environment and project being evaluated. With careful consideration of the weights, and by turning rules on and off for each model run, it is possible to separate model variability from environmental variability.

By default, each rule has a weight of [1.0]. By specifying a weight less than [1.0] the truth of the antecedent is reduced by that proportion. For example, specifying the rule in Section 11.2.2.1.2 on page 138 with a weight

```
(Weight=0.8) IF dissolved_oxygen is slightly high,
    THEN egg_hatch is reduced.
```

multiplies the truth (degree of membership) of the antecedent *dissolved_oxygen is slightly high* by [0.8]. This reduces the influence of this rule by 20 percent. Similarly, entire policies (or rule blocks) can be weighted so that they have difference influences on the solution fuzzy set and the crisp output.

11.3 Significance, Acceptability, Sustainability

These concepts are at the core of environmental impact assessments and are the basis upon which regulatory and policy decisions are made. The meaning of each of these terms is inherently uncertain, imprecise, and fuzzy. Each of these words represents a concept that is not directly related to a measurement of size, time or place (such as "large," "recent," or "near"); that is, each is a Type-2 fuzzy variable (Section 9.6 on page 85). While each term describes a word concept without a basis in a measurement, they are different from each other in how they are derived and used with a fuzzy system environmental impact assessment model. Further, each concept can be redefined within the social, judicial, political, regulatory, and geographic climate in which the assessment is conducted while retaining its power of explanation and support of decisions.

11.3.1 Significance

Significance is a solution Type-2 fuzzy set. It is derived from evaluation of impact type, direction, magnitude, duration, ærial extent, and other sources of this impact type within the evaluation area.

Chapter 6 divided impact evaluation into three components: identifying impacts by name and direction (positive or negative), measuring the relative size of the impact and evaluating the significance of that magnitude of impact on the socioeconomic and environmental components of greatest interest for a particular project. Identification of impact names and directions is comparatively straightforward for those who understand the structure and function of the systems. It is also relatively easy to gain agreement of the impact identities and directions among decision-makers, technical staff, and other interested parties. Describing impact magnitude may also be relatively simple when based on scientifically and economically accepted measurements. Otherwise it can be qualitatively described. Determining impact significance is a process that may generate disagreement and heated discussion when done subjectively. Quantifying the concept of *significance* involves a rule policy consisting of several antecedent fuzzy sets, each appropriate to a different type of input data and measurement criterion. The inference mechanism must be chosen with care and rules must have appropriate, and different, importance weights.

Because significance determination is central to environmental impact assessments the process will be described in detail. While the following is a generic description that will work in most—if not all—situations, it can be modified to better suit particular needs.

The components included in the significance determination are derived from the list in Section 6.2 on page 43—

- Likelihood of occurrence
- Direction and magnitude
- Areal extent or distribution

- Duration
- Impact reversibility
- Whether impact can be mitigated; if so, to what degree
- Timing of occurrence relative to project phases
- Geographic impact scale

The significance policy (rule block) is organized as shown in Figure 11.4 on the following page. The linguistic variables (with their fuzzy term sets) are designed to capture the underlying meaning in each of the listed criteria (Figures 11.5 and 11.6). The details of fuzzy set shape and the Universe of Discourse must be adjusted for each assessment and location. For all linguistic variables displayed in Figures11.5 and 11.6 on page 154, an α-cut of [0.2] is applied. That is, any degree of membership ($\mu[x]$) less than [0.2] does not contribute to the solution. This threshold is used to ensure that the significance determination has real-world meaning and is not just a number calculated to have a numeric output. While importance weights are assigned to rules, not linguistic variables, there is value in considering the weights that might be applied to each criterion at this stage. There are no guidelines; each project and location is different from every other one.

11.3.1.1 Likelihood of Occurrence

Likelihood of occurrence (Figure 11.5(a)) is part of the impact identification process. Not all activities of a given type will impact the environment, economy or social structure to the same degree (if at all). The likelihood of impacts to birds from rotating blades of turbines at a wind farm will depend on factors such as whether the wind farm is located along a migratory route, whether the turbines will be run continuously or only during periods of peak loads (diurnal and seasonal variations, and the visibility of the blades to flying birds. The likelihood of water quality degradation for a marine terminal might be quite high, in fact, almost certain because accidental spills and normal operations will result in discharge of undesired chemicals in near-shore waters. In terms of impact significance, the likelihood of occurrence contributes to the final determination.

The linguistic variable has five fuzzy sets across the Universe of Discourse that can be considered to be probability. The term sets are

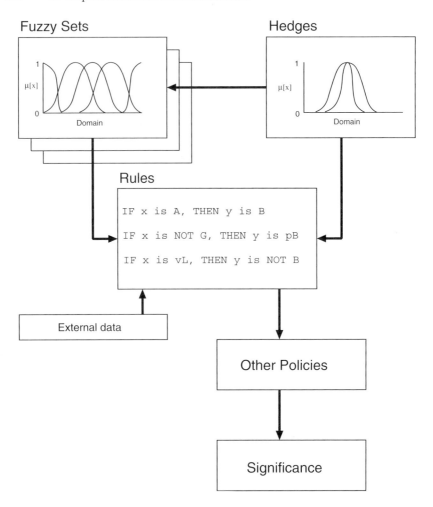

Fig. 11.4. The overall structure of the significance policy.

labeled *Rare, Uncommon, Common, Frequent,* and *Certain.* The three interior fuzzy sets are trapezoidal, while the two exterior fuzzy sets are shouldered. The relationship of the shape to the underlying semantic meaning of likelihood of occurrence is that only an estimate of likelihood can be made and that estimate is equally true for a range of values along the Universe of Discourse. For values within that range,

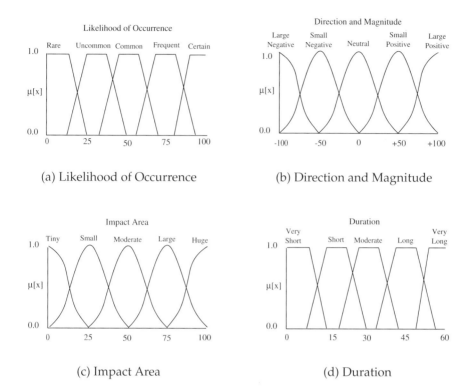

Fig. 11.5. Fuzzy sets for determining impact significance: a) likelihood of occurrence; b) direction and magnitude; c) aerial extent; d) duration.

the degree of membership (the truth value) is [1.0]. Values less than [1.0] change rapidly and evenly during the transition from one term to the next.

11.3.1.2 Direction and Magnitude

Direction and magnitude of an impact (Figure 11.5(b)) are considered together because the significance value is a factor of both simultaneously. For some impacts this criterion is determined by measurement—e.g., clearing all vegetation from a parcel of land results in a complete

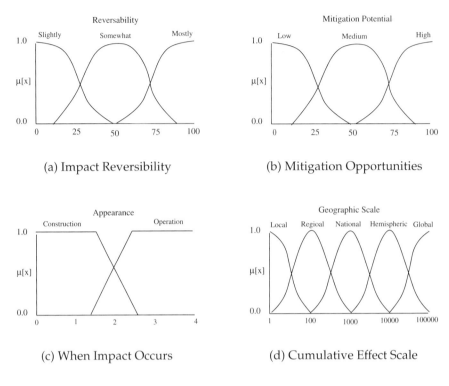

Fig. 11.6. Fuzzy sets for determining impact significance: a) reversibility; b) mitigation; c) occurrence period; d) geographic scale

loss of vegetation, so the direction and magnitude of impact can justi-fiably be set to the maximum negative value. On the same assessment, planned changes in land use along a stream may be designed to add an additional 50 percent to salmonid spawning and rearing habitats along the reach. The additional high-quality habitats would be a positive im-pact of a defined amount.

The Universe of Discourse has the range [–100,+100] and is ex-pressed by five fuzzy term sets labeled *Large-negative*, *Small_negative*, *Neutral*, *Small_positive*, and *Large_positive*. The term sets are sigmoid in shape because change from one term to another is gradual and minor value differences near $\mu[1.0]$ change the degree of membership slowly. The number of term sets can be set to whatever is most appropriate for a particular assessment, and the Universe of Discourse can also be altered to fit the situation. However, as with all the criteria of significance, the same linguistic variable definition must be used for all impacts within a given environmental impact assessment.

11.3.1.3 Areal Extent

The areal extent of the impact (Figure 11.5(c)) is relative to the project being assessed. If the site is away from an urban area, then the economic impact area may be relatively small because jobs and other spending may be concentrated in the nearest town. A seaport in a metropolitan area may have much larger, regional economic impacts. Air quality impacts might not be discernible beyond the project boundary, and wetland loss may be only about 3 hectares. Perhaps the site has a single nest for a listed bird species so the impact of the project on that one nest sight might be considered to be large. This criterion defines the direct impact size relative to the project size.

The Universe of Discourse is [0, 100] simply for illustration. This range should be selected based on the size of the assessment area with a buffer zone on the outside. The units should be appropriate measures of area: acres, hectares, kilometers-squared, or whatever is appropriate. In most environmental impact assessments it is not useful to describe an area exactly because impacts are relative to each other and the degree of significance is based on relative contributions of all criteria. Measures are converted to fuzzy degrees of membership in the variable's term set of *Tiny*, *Small*, *Moderate*, *Large*, and *Huge*. These linguistic terms also accommodate the areal range of all the components included in the assessment.

11.3.1.4 Duration

In addition to the intensity, direction and direct area impacted, the length of time the impact exists contributes to the overall significance. Figure 11.5(d) illustrates how the length of time the impact is in force can be converted from a number of years to a degree of membership in a fuzzy set ranging from *very_short* to *very_long*.

Other aspects of duration may be important to evaluate impact significance. These would require different—or additional—fuzzy term sets. For example, a hydroelectric dam may impact migratory salmonids only when the turbines are operating. At other times, water is spilled from the reservoir behind the dam to the tail race below the dam. Fish can pass in both directions (assuming there is a fish ladder, of course) without passage through the spinning turbines. In this case, the frequency of operation would contribute to the significance of impacts. A peaking plant (run only when there is very heavy demand on the electrical system) has a different operating schedule from a continuous operations plant.

The Universe of Discourse is set for the appropriate period, from days to decades. Five fuzzy term sets is appropriate for this linguistic variable; three is too coarse and seven is probably too fine for times that can only be estimated or forecast. Trapezoidal shape is easy to interpret with regard to a span of time: for each of the fuzzy sets full membership is observed over a period and the degree of membership (truth value) decreases constantly and linearly on either side of that plateau. Intuitively, one would not expect the impact to vary in a sigmoid manner over a defined period of time.

11.3.1.5 Reversibility

Some impacts are temporary, while others are permanent. Conversion of an agricultural field to a residential subdivision adds impervious area as roads, driveways, and rooftops (permanent change); but loss of vegetation and wildlife habitat could be replaced by lawns and trees (temporary change). Construction of a large electrical generating facility will increase the local economy (jobs and spending), but the level may be reduced when the operational phase begins. The permanence and degree of reversibility of an impact are factors in its significance. Roads are created when timber is harvested, but the roads will be taken out by grading and revegetation when logging and replanting is finished.

Any change to a natural system cannot be reversed in the sense of being completely undone, so the system is totally indistinguishable in structure and function as it was before. However, from a practical perspective, functional reversal can certainly be achieved. It may take time (as is the case with plant succession after a fire, windstorm, volcanic eruption, or commercial harvest), but natural processes will drive change toward what the system was before. The degree of reversibility is subject to evaluation by value systems as well as by tape measure or scale. That is why it is represented in the environmental impact assessment system model as having three fuzzy term sets: *Slightly*, *Somewhat*, and *Mostly* (Figure 11.6(a)). Every point of view can be captured within this range for evaluation of an impact's reversibility without causing heated argument and gridlock over what value to assign. The Universe of Discourse represents proportion or percentage. Because wide, gradually changing sigmoid curves are used to represent the range of values, the resolution is deliberately retained as coarse. This helps to keep this factor in perspective with the other factors contributing to the degree of significance of an impact. The relative contribution of reversibility as a factor is reflected in the weight assigned to it. There are undoubtedly assessments where the degree of reversibility of an impact (whether positive or negative) is of great importance to the decision-makers. In other situations it will be comparatively less important in determining significance. A coarse resolution of both measurement and expression of belief or value system facilitates the appropriate contribution of this factor.

11.3.1.6 Mitigation Degree

A wetland may be filled during the construction phase of a project, but another one of equivalent or better size, function, and values could be created to replace it. In a number of jurisdictions, mitigation ratios are established by regulation with the goal of reducing the significance of impacts on certain environmental components. Mitigation, or replacement of functions or values lost, may occur on-site or off-site. Compensation for a negative impact affects significance in a complex manner. A consideration of mitigation degree and its influence on impact significance is when the mitigation is developed and is functional compared with the impairment or loss incurred. Replacement of wetlands after a project is functionally completed (e.g., when a mineral resource is mined out) *may* have different significance than the creation of new housing, roads and schools prior to project development.

The large number of factors included in determining the potential for mitigation is best quantified by a linguistic variable having only three, broad fuzzy term sets: *Low*, *Medium*, *High* (Figure 11.6(b)). If the Universe of Discourse is considered the percentage of the impact that can (or will) be mitigated, then that value can be converted to a degree of membership in one or more of the fuzzy term sets. In addition, differing opinions—including those of conflicting experts—can be included in the analyses.

This linguistic variable can be structured to include those factors that are deemed most appropriate for inclusion in the assessment. Factors include the extent of the potential or actual mitigation, when it is applied relative to the project time schedule, the effectiveness of mitigation against some standard and the relative ease of monitoring mitigation progress toward design goals.

11.3.1.7 Occurrence Timing

When the impact occurs relative to a project schedule affects its significance. The time of occurrence is different from the duration of the occurrence and has a different effect on significance. However, there is an interaction between the two. One impact might occur at the very beginning of a project and last for a varying period. Its significance can be quite different than the same impact's occurring toward the end of a project but with the same duration. Other impacts could repeat either regularly or irregularly over the course of the project. Depending on the other characteristics of the impact, time of occurrence can have varying influence on the impact's significance.

In the model being described, the decision was made to divide the linguistic variable into two fuzzy term sets, *Construction* and *Operations* (Figure 11.6(c)). This might be the appropriate way to represent the timing occurrence of fish passage impacts when a hydroelectric dam—with fish ladder and juvenile by-pass facilities—is being assessed. During construction passage of fish up- and downriver will be interrupted. However, assuming that the fish passage facilities operate as intended and designed, the impact should diminish during the operational phase of the dam.

Obviously, the fuzzy term set must be designed to reflect the type of project and expected impacts so that it well represents the underlying semantics of when an impact occurs relative to the project timeline.

11.3.1.8 Geographic Scale

The geographic scale of an impact is different from the areal extent of the impact; the latter is direct while the former is indirect. For example, the loss of a 5-hectare wetland in a semi-arid setting has a comparatively small local areal extent. However, if it is the only wetland in a region several hundred square kilometers in size, its significance will be different than if the local area has several other wetlands. The geographic scale influence on impact significance can be used as a measure of cumulative effects. This is the variable that considers the broader setting in which the impact occurs and affects the calculation of overall significance.

The linguistic variable of Geographic_scale is partitioned into five fuzzy term sets: *Local*, *Regional*, *National*, *Hemispheric* and *Global* (Figure 11.6(d)). The Universe of Discourse in the illustrated example is square kilometers, but the range, number, and placement of term sets and their labels need to be adjusted to the specific assessment being conducted.

The geographic scale of impact is not always negative. In the case of a marine shipping terminal, goods from hundreds of kilometers away have access to world markets. Similarly, products from other countries can be distributed and sold for lower prices when the point of entry to the receiving country is close to the point of final sale. In some environmental impact assessments, for example, the evaluation of existing conditions there are no project-specific impacts but there are effects and implications for environmental, economic and social conditions beyond the area of immediate attention. To that extent, significance can be determined because it provides more insight to the decision-makers and can result in more informed decisions.

11.3.2 Acceptability

Acceptability is more abstract as a concept than is significance but it can also be narrowly defined and easily evaluated. In the most simple case, environmental acceptability occurs when no statutory or regulatory standard is exceeded. That is, water quality is not degraded below the point where it is considered suitable for human consumption, food crop irrigation, or use by livestock and wildlife. Similarly, air quality remains within defined limits for particulate matter, acid precipitation precursors, and ozone-depleting chemicals. When there are no toxic effects on nontarget organisms or the creation of hazardous wastes, an activity can be considered acceptable. There are many instances of this straighforward definition and application of acceptability. There are also many more complex situations that require more indirect methods to resolve.

Environmental impact assessments generally do not explicitly evaluate acceptability. The criterion being evaluated is significance, and the decision is assumed to be acceptable when the alternative with the least significant negative impacts is selected. However, since the selection among all evaluated alternatives usually is based on political values (the environmental, economic, and social having been considered in the scores assigned to each alternative and presented to the decision-makers), it can safely be assumed that the choice meets various criteria of acceptability. That decision, however, is not under the control of those conducting the assessment in many situations so it will not be considered in detail here.

When there is reason to incorporate a formal evaluation of acceptability, it is most appropriately described as a Type-2 fuzzy set. Unlike the subjective term, *significance,* which is a consequent of antecedent, measurable terms, *acceptable* (or *acceptability*) cannot be derived by other components. The definition of what is acceptable is based solely on individual and group values. Creating such a fuzzy set requires a method that captures the inherent uncertainties of meaning in a useful way. Such a method was developed by Mendel [20] to answer the question, What is the smallest number of words (or phrases) that covers the interval 0–10? His method is applied to the creation of a fuzzy set for *Acceptability*. These results can be used as they are, or a new fuzzy set representing local values and beliefs can be created.

For consistency, Mendel's set of 16 labels will be used; their order of listing does not suggest a uniform increase or decrease in value—

- None
- Some
- A good amount
- An extreme amount
- A substantial amount
- A maximum amount
- A fair amount
- A moderate amount
- A large amount
- A small amount
- Very little
- A lot
- A sizable amount
- A bit
- A considerable amount
- A little bit

In Mendel's situation university students were randomly chosen and asked to assign a starting and ending value in the range [0,10] to each term with overlap among ranges allowed. For purposes of an environmental impact assessment the participants in the creation of this fuzzy set should be as large a number of stakeholders as possible. The instructions given to each survey participant are those of Mendel [20, page 71]:

> Below are a number of labels that describe an interval or a "range" that falls somewhere between 0 to 10. For each label, please tell us where this range would start and where it would stop. (In other words, please tell us how much of the distance from 0 to 10 this range would cover.) For example, the range "quite a bit" might start at 6 and end at 8. It is important to note that *not all the ranges are the same size.*

Some respondents will use only integers, while others will provide decimal values as ranges for each label. This variety and individual expression is to be encouraged because it best captures the variability in how individuals think of the term *acceptable.*

Each label is given a start value and an end value. From these can be calculated the mean and standard deviation for the starting and ending values of each label. When the results are plotted, the range between the average start and average end values for each label define their interval along the Universe of Discourse while the standard deviations represent uncertainties at each end (Figure 11.7 on the following page). This figure shows that there is a lot of overlap in the range covered by similar terms. Therefore, simplification is in order.

Close examination of the intervals and uncertainties (i.e., the distance between the extremes of the standard deviations) indicates that with this data set three labels can cover the range [0,10]: *none to very little, a moderate amount,* and *a maximum amount.* However, this extreme reduction does not provide the more desirable granularity that can be achieved with five labels across the range: *none to very little, some, a moderate amount, a large amount,* and *a maximum amount* (Figure 11.8 on the next page).

In Section 9.6 on page 85 a Type-2 fuzzy set was described as having each value along the Universe of Discourse be a Type-1 fuzzy set that represents the uncertainty about the truth value (degree of membership). Figure 11.7 on the next page can be converted into a Type-2 fuzzy set membership function by having each term span the interval (shown as the thick horizontal line) and the footprint of uncertainty be

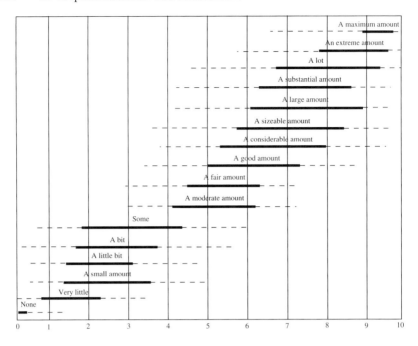

Fig. 11.7. All labels with interval and uncertainty. From [20, page 74].

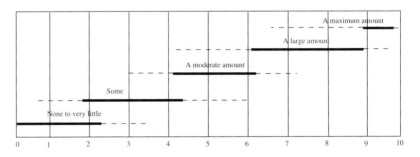

Fig. 11.8. The final set of labels for *Acceptability*. Each will be represented as a Type-2 fuzzy term.

the expression of the standard deviations. To illustrate this transition, Figure 11.9 shows how the label, *A moderate amount*, is converted into the fuzzy set, *Moderate_amount*. Also shown on this figure are the degrees of membership of two values along the support set of this membership curve. The created Type-2 fuzzy set, *Moderate_amount*, is of the

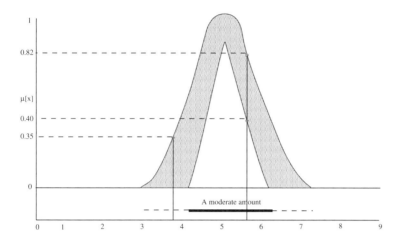

Fig. 11.9. Converting the label *A moderate amount* into the Type-2 fuzzy set *Moderate_amount*. Also shown are the degrees of membership for two specific values.

uncertain mean type. The standard deviation is known (shown by the dashed lines), but the mean value is somewhere in the range of the heavy line.

At the x value of 3.8 the degree of membership $\mu[3.8] = 0.35$; therefore, this x value has the truth of 0.35 in the fuzzy set *Moderate_amount* of acceptability. This degree of membership is a single value because $x = 3.8$ falls within the footprint of uncertainty (FOU) of the Type-2 fuzzy set. At a higher value for x, 5.75, the degree of membership is somewhere between the values $\mu[5.75] = [0.40, 0.82]$. In other words, the truth of this value belonging to the fuzzy set is somewhere in the range [0.40, 0.82]. Using the mean value, $\mu[5.75] = 0.61$, will suffice for many applications, but the uncertainty can be applied to a fuzzy system model to produce a more precise result.

This process began by asking people to partition a numeric range (0, 10) into smaller ranges represented by an initial set of 16 labels. Because there are no measurement references involved, the results of the survey (mean and standard deviation) for both the starting and ending values for each subrange represent belief systems and values applied to the meaning of those labels. The set of 16 labels have a lot of overlap and can be reduced to a set of five subranges that span the interval [0,10]. The interval is assigned as the Universe of Discourse for the linguistic variable, *Acceptability*, and the five subintervals are the term set for that variable. Each of the five labels is converted to a Type-2 fuzzy set that reflects both values and uncertainties associated with each term. The result is a numeric value for *Acceptable* (or *Acceptability*) that can be used in a fuzzy system environmental impact assessment model or any other fuzzy system model where this linguistic variable is a component.

11.3.3 Sustainability

The United Nations Division for Sustainable Development defines sustainability as "development that meets the needs of the present without compromising the ability of future generations to meet their own needs" [11]. Sustainability is similar to significance as a solution Type-2 fuzzy set in that it is derived from antecedent measurements that include time, economic factors, societal value factors, and resource factors. Where sustainability differs markedly from significance is in the number and measure of the antecedent components from which it is calculated. Because sustainability means different things to different people in various regions and countries, there is no single set of antecedent linguistic variables that will fulfill all needs.

In general, the goals of sustainability planning and actions are to provide long-term benefits from the extraction of natural resources to all parties involved. Communities that are asked to make changes to accommodate larger populations, heavier traffic, or removal of resources for the benefit of others want to know that the jobs, education, health care, and other socioeconomic benefits can continue after the resource is locally depleted.

Whether sustainability is considered a process or a goal, it is a highly emotional, complex concept built strictly on values and priorities of stakeholders that may be at odds with each other. This is precisely why the methods and approaches of fuzzy system models are so well suited to reaching consensus.

The process should begin with a list of all possible definitions of "sustainable" or "sustainability" and a list of what components are part of the concept. These can be ranked using the same process used in scoping an environmental impact assessment. The most important components can be considered as linguistic variables and have fuzzy term sets developed for them. All stakeholders can participate in determining acceptable shapes, support set ranges, and overlaps so as to produce the model core that reflects the range of local values and beliefs. Almost always a compensatory aggregation operator will be used to infer the consequent fuzzy set from the antecedent fuzzy sets. This is because the combination of AND and OR allows a critical middle ground between meeting all criteria (AND) or any one of the criteria (OR).

What may turn out to be the most arduous part of the process is determining the rules to be applied. There are probably no "experts" who can be interviewed by a knowledge engineer to elicit how decisions are made under various input conditions. So the appropriate approach is to collect opinions and feelings from the stakeholders and vote on the results. Of course, with suitable modeling tools real-time sensitivity analyses ("What if ... ?") can be conducted.

There is no way to know now just how sustainable any policy will be at specific times in the future. However, when we incorporate the thinking of everyone with an opinion into the process the probability of developing cooperation is increased.

Multi-Objective, Multi-Criteria Decision-Making

Objectives are goals or outcomes. In environmental impact assessments there are always several objectives, sometimes conflicting with, other times complementing, each other. Development projects have objectives of profits or services. Considerations of impacts on the natural environment and the socioeconomic setting in which the project is located are the criteria by which the various objectives are evaluated. Since the effects of each criterion will be different on each objective, the decision-making process must involve all objectives and all criteria. A hydroelectric dam may have objectives of producing electrical power, flood control, irrigation water supply, and recreation. A gravel pit on farmland may have the objectives of both extracting aggregate and returning the land to the farmer without rocks and with optimal farming contours. A seaport may have objectives of maximizing import and export of certain commodity types, operating a ship building and repair profit center, leasing buildings and facilities for industrial and commercial operations, and running a gas-turbine, cogeneration power plant to provide electric power, heat, and process steam for all facilities. In many cases, perhaps most, the objectives have different priorities for separate groups of stakeholders. None of the objectives for the representative projects listed above included any of the components of an environmental impact assessment.

Criteria can be considered constraints that must be satisfied when an objective is reached. Criteria for a project could be included in the descriptions of objectives: maximize profit, minimize costs, constrain environmental disturbance to the minimum acceptable level, generate more jobs. Each criterion must be satisfied regardless of which objective is accepted.

In every real world situation there are objectives and criteria along continua between extreme positions. However, without well-structured, solidly based, objective (compared with subjective) procedures for finding the most acceptable project goals and threshold of criteria satisfaction, the process may result in a zero-sum game. For one side to win, the other side must lose. While this is completely satisfactory to the winner, it leaves innocent bystanders out of the decision-making process and it creates very unhappy losers.

Among the many possible methods proposed and used in different multi-objective, multi-criteria decision-making processes, one that could be used in environmental impact assessments is based on the Analytic Hierarchy Process (AHP) developed by Thomas L. Saaty in 1970 [29]. Figure 12.1 is a visual model of a hierarchy for port development. The goal is the optimal development path, the objectives that are available to reach that goal are maritime operations, heavy or light industrial activities or commercial leases. The constraints on achieving those objectives are profits, jobs, wetlands, wildlife habitats, and fish habitats.

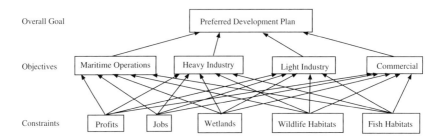

Fig. 12.1. Example of an AHP for seaport development. Four objectives and three criteria (constraints) are shown as levels in the decision-making hierarchy.

Saaty developed the AHP as both a theory and method for "modeling unstructured problems in the economic, social and management sciences" [29, page 3]. Economic theory has been developed around the concept of money as the unit of measurement. There are well-established approaches and formulas for valuing goods and services that work so well they permit people to function in the economic decision-making process. However, money is not as well established as the measurement unit for social or political values. Social values are the basis for assessing environmental impacts. While the explicit dollar values of æsthetics, recreation, open space, and other social values may not always be expressed by environmental impact assessment decision-makers, they are probably present if only at a subconscious level. This is to be expected because decisions are based on trade-offs among various alternatives, and there must be a common denominator upon which to make comparisons. If an alternative to each individual's monetary amount could be applied to the valuation of trade-offs among social beliefs, then there would be a measurement system that could be universally accepted. One such alternative is by comparing objects, concepts or values by pairs.

To illustrate how this process works, consider the construction of a dam in an economically developing region. There are four potential functions of the dam: hydroelectric power generation, flood control, irrigation of agricultural crops, and commercial fish farming. The planners have determined that cultural alterations, economic benefits, environmental impact potential, and transportation distances are the most important criteria in evaluating each function. The planners need a way of formalizing the relative importance of each criterion and evaluating each alternative function relative to those criteria.

The first step in applying the AHP to the planners' problem is to determine the relative importance values of each criterion. This is done by comparing the four criteria pair-by-pair using the relative importance scale in Table 11.1 on page 130. The results of their evaluation is shown in Table 12.1 on the following page. These values represent the relative importance of each pair-wise comparison. The table is symmetrical because each self-comparison is equal to 1 and equivalent cells are reciprocal: that is, $a_{ij} = k$ and $a_{ji} = \frac{1}{k}$. Once this matrix is established, the next step is to compute a vector of priorities. The exact solution is presented in the following section; here an estimate is made using one of Saaty's "good" methods [29, page 19]:

> Divide the elements of each column by the sum of that column (i.e., normalize the column) and then add the elements in each

Table 12.1. Planners preferences on dam evaluation criteria (objectives).

	Culture	Economy	Environment	Transportation
Culture	1	1/5	1/3	1/2
Economy	5	1	2	4
Environment	3	1/2	1	3
Transportation	2	1/4	1/3	1

resulting row and divide the sum by the number of elements in the row. This is the process of averaging over the normalized columns.

Doing these calculations for the values in Table 12.1 yields the results in Table 12.2. Looking at the "Average" column in this table it appears

Table 12.2. The weights of the function evaluation criteria.

	Culture	Economy	Environment	Transportation	Average
Culture	0.091	0.102	0.091	0.059	0.086
Economy	0.455	0.513	0.545	0.471	0.496
Environment	0.273	0.256	0.273	0.353	0.289
Transportation	0.182	0.128	0.091	0.118	0.130

that approximately half the planners' objective weight is on economic benefits, approximately 30 percent on environmental impacts, 13 percent on transportation distances, and about 9 percent on potential for cultural alterations.

To understand why this makes sense, look at the values in the first column of Table 12.1. The value of Culture has been set to 1 and the other criteria normalized relative to that value. Each of the other columns is normalized with regard to Economy, Environment, and Transportation, respectively. If all the planners were perfectly consistent each column would be identical, except for the normalization. If we divide the value in each cell by the column's total, all columns would be the same with the cells reflecting each row's relative weight. The average across all row values corrects for minor inconsistencies among the planners

Once the criteria have been weighted, each goal (that is, dam function) is evaluated for each criterion. The same two-step process of evaluating goals, then normalizing each column and averaging each normalized row, is followed for each goal. For the Culture criterion the planners evaluate each dam function as shown in Table 12.3. The

Table 12.3. The relative importances of each dam function on cultural alterations.

	Electrical	Floods	Irrigation	Fishing
Electrical	1	6	6	1/9
Floods	1/6	1	1/3	1/3
Irrigation	1/6	3	1	1/3
Fishing	9	3	3	1

weighted values of each dam function on potential cultural alterations is in Table 12.4; almost half of all potential cultural alterations would result from being able to fish in the reservoir, about 30 percent from available electricity, 13 percent from irrigated crop lands and less than 8 percent from flood control.

Table 12.4. The relative effects of each dam function on potential cultural alteration of the local population.

	Electrical	Floods	Irrigation	Fishing	Average
Electrical	0.097	0.462	0.561	0.062	0.300
Floods	0.016	0.077	0.032	0.188	0.078
Irrigation	0.016	0.231	0.097	0.188	0.133
Fishing	0.871	0.231	0.290	0.562	0.489

The next criterion (objective in AHP and constraint in decision science) is economic benefits. These might include the creation of jobs, increased exports with decreased imports, reduced subsidies from the central or local governments to residents, and an increase in general health of the population. The planners determine that the four dam goals would have relative economic impacts presented in Table 12.5.

Table 12.5. The relative importances of each dam function on the economy.

	Electrical	Floods	Irrigation	Fishing
Electrical	1	7	5	3
Floods	1/7	1	1	1/5
Irrigation	1/5	1	1	1/3
Fishing	1/3	5	3	1

After averaging the rows on normalized columns, the relative importance of the four dam functions on the economy is shown in Table 12.6.

Table 12.6. The relative effects of each dam function on the economy.

	Electrical	Floods	Irrigation	Fishing	Average
Electrical	0.597	0.500	0.500	0.662	0.565
Floods	0.085	0.071	0.100	0.044	0.075
Irrigation	0.119	0.071	0.100	0.074	0.091
Fishing	0.199	0.357	0.300	0.221	0.269

For this criterion, approximately 56 percent of economic benefit accrues from hydroelectric generation at the dam, approximately 27 percent from the fish farming industry, 9 percent from irrigated agriculture, and 7.5 percent from flood control.

Environmental impacts from each of the four dam functions are considered by the planners. They determine the relative importances shown in Table 12.7. Almost 42 percent of the anticipated environmen-

Table 12.7. The relative environmental impacts of each dam function.

	Electrical	Floods	Irrigation	Fishing
Electrical	1	1/7	1/5	3
Floods	7	1	1/3	1/6
Irrigation	5	3	1	1/7
Fishing	1/3	6	7	1

tal impacts are expected to come from the fish farms, just over 20 percent each from irrigating farm lands and hydroelectrical power generation, with the remaining 17.5 percent associated with flood control (Table 12.8).

Table 12.8. The relative environmental impacts of each dam function.

	Electrical	Floods	Irrigation	Fishing	Average
Electrical	0.075	0.014	0.023	0.696	0.202
Floods	0.525	0.099	0.039	0.039	0.175
Irrigation	0.375	0.296	0.117	0.033	0.205
Fishing	0.025	0.592	0.820	0.232	0.417

The final criterion, transportation distance, is important to the planners because of the costs associated with road and electrical transmission line construction and maintenance, and the time required to get fresh produce and fish to markets or shipping ports for export. Traffic volumes, fuel consumption, air pollution, and vehicular safety concerns also fall within the establishment of this objective's relative importances associated with each of the four dam goals. Taking all of these factors into consideration, they came up with the relative importances shown in Table 12.9 . For this criterion, electrical transmission

Table 12.9. The relative transportation distance importance of each dam function.

	Electrical	Floods	Irrigation	Fishing
Electrical	1	9	5	3
Floods	1/9	1	1/7	1/6
Irrigation	1/5	7	1	1/4
Fishing	1/3	6	4	1

distance accounts for about 54 percent of importance and is followed by fish farming (27 percent), irrigation (14.5 percent, and flood control (4.5 percent). These values are in Table 12.10 on the following page.

Table 12.10. The relative importances of transportation distance on each dam function.

	Electrical	Floods	Irrigation	Fishing	Average
Electrical	0.608	0.391	0.493	0.679	0.543
Floods	0.068	0.043	0.014	0.038	0.041
Irrigation	0.122	0.304	0.099	0.057	0.145
Fishing	0.203	0.261	0.314	0.226	0.271

The relative weights of each criterion for each dam function are summarized in Table 12.11. Now the values for each goal (a dam func-

Table 12.11. Relative criteria scores for each dam function.

	Electrical	Floods	Irrigation	Fishing
Cultural	0.300	0.078	0.133	0.489
Economic	0.565	0.075	0.091	0.269
Environmental	0.202	0.175	0.205	0.417
Transportation	0.543	0.041	0.145	0.271

tion) can be calculated for each constraint (criterion). The value of hydroelectric power generation is

$$(0.300)(0.086)+(0.565)(0.496)+(0.202)(0.289)+(0.543)((0.130) = 0.435 \tag{12.1}$$

for flood control the value is

$$(0.078)0.086)+(0.075)(0.496)+(0.175)(0.289)+(0.041)(0.130) = 0.100 \tag{12.2}$$

for agricultural irrigation the value is

$$(0.133)(0.086)+(0.091)(0.496)+(0.205)(0.289)+(0.145)(0.130) = 0.135, \tag{12.3}$$

and for fish farming in the reservoir the value is

$$(0.489)(0.086)+(0.269)(0.496)+(0.417)(0.289)+(0.271)(0.130) = 0.331 \tag{12.4}$$

Based on the planner's decisions, hydroelectric power generation at the dam has the highest overall value among the four benefits the project provides. Fish farming in the reservoir created behind the dam has the second-highest value and is followed by agricultural irrigation and flood control, in that order. More importantly, there is an easy-to-follow justification for the results and it is very easy to explore the results if different comparative values are made among objectives and criteria. The AHP is one multi-objective, multi-criteria decision-making process that eliminates decisions' being made arbitrarily or capriciously.

A limitation of the AHP in environmental impact assessments is the assumption that each objective (dam function, in the example here) has a weight equal to the others. In most development projects one objective (or goal) has a higher value than does another objective. Consider the dam example—there are four criteria (also referred to in the literature as objectives or constraints), and the AHP analysis was conducted on the implicit assumption that the four criteria have equal weight. Yager [39] multiplied the resultant weights (the normalized eigenvector) by the order of the matrix, so the criteria are weighted rather than being equal. This is a more appropriate approach when values are subjective (i.e., fuzzy) and have different importance values to stakeholders. To clarify why this is important, think about the considerations of the planners involved in the dam decision. They might decide that it is worth some economic and cultural values to decrease environmental impacts. Or, they might decide that economic benefits are worth some environmental impacts and changes to cultural practices. Different objective values are probably more common in real world multi-criteria decision-making than are equal values for each one. Several other authors take similar approaches to multi-objective decision-making, multi-criteria decision-making or both, e.g., [24, 40, 44]. There are also differences in the scales used to measure criteria and goals. In the example above linguistic terms (an ordinal scale) were used because this is most easily understood by decision-makers. But, the linguistic values in Table 11.1 on page 130 are converted to cardinal numbers (a ratio scale) that can be manipulated in a matrix.

An application of the AHP was used to develop environmental quality indices for large, industrial development alternatives along the east coast of Iceland [34]. The results illustrate both the potential and the limitations of this method for handling the complexities of environmental impact assessments.

Two alternatives were a large aluminum smelter powered by a hydroelectric dam to be built on the edge of a glacier and petroleum refinery to process crude oil transported to a coastal city by ship from arctic Russia. A third alternative was contrived; this "green" option consisted of general development, eco-tourism and knowledge-based industry.

Sólnes defined an environmental quality index as the weighted sum of selected environmental and socioeconomic factors, and he asked five colleagues to join him in providing the values as input to the AHP. The reported final results were that the two real industrial projects had environmental quality indices well below that of the third alternative. The complexities of computing the values of many matrices lead the author to reduce the considered factors to a manageable few. It is highly likely that this compaction and the few, collegial university faculty who provided the values as baseline data produced results different from those of a more extensive evaluation.

The decision hierarchy was defined on four levels. The top level (the goal) was the project evaluation, the environmental quality index. The next-lower level (the objectives) had four nodes: technical assessment, economic impacts, environmental impacts, and social impacts. The third level (the attributes) contained six leaves for technical, five leaves for economic, five leaves and a node for environmental, and six leaves for social issues. The lowest level of the hierarchy held four leaves in the industrial pollution attribute of the environmental objective. The analysis, however, combined all economic attributes into a single factor and all the social attributes into a second factor. The six environmental attributes on level three have only two (land use and water systems) combined. The "population" of stakeholders included the author, four engineers, an economist and a social scientist.

This article is an illustration of how the AHP could be used to support decisions based on an environmental impact assessment. But the computational effort required a well-justified combination of factors and limited population of stake holders, and this restricts its applicability in support of a decision. Another consideration in selecting a technically sound and legally defensible model for multi-objective, multi-criteria decision support is the ability to quantify subjectivity at a fine scale and manipulate subtle differences in a mathematically rigorous manner. Fuzzy sets and fuzzy logic overcome the limitations of the AHP (and its enhanced version, the Analytic Network Process or ANP).

Part III

Application

13

Introduction

13.1 What This Part Includes

Part II explains why fuzzy sets and fuzzy logic overcome the inherent limitations of the traditional approach to environmental impact assessments. This part demonstrates how the modern approach works. The example is based on an environmental impact assessment conducted under the rules of the Washington State Environmental Policy Act (SEPA) by a large team lead by the JD White Co. [2].[1] The actual environmental impact assessment was done the traditional way as described in Part I. Much of the information required to apply the fuzzy logic approach in an actual assessment is missing from the written report.

Because the example was not designed or carried out based on requirements of the modern approach to a real environmental impact assessment, liberties are taken to generate required missing data and not all components are included in this demonstration model. However, using a real situation as a base is better than using a completely contrived situation.

There are many expert system shells available as proprietary or open source distributions. Some are useful for fuzzy logic and others are not. There are also tools, such as MatLab®, SciLab, and Octave (the

---

[1] The author of this book did not participate in any aspect of the SEPA process. All information for this example is extracted from the written document made available to the public.

latter two are open source equivalents of the first application) that are general mathematical and simulation solution packages. All of these have fuzzy logic components available. However, the software used in the example in this part is a proprietary system called FuzzyEI-Assessor™.

13.2 Description of the Project

The Port of Vancouver (USA) occupies the north bank of the Columbia River between the Interstate 5 bridge and a state wildlife refuge several miles to the west (Figure 13.1 on the next page). The Columbia Gateway area consists of 1,094 acres of land, designated by the port as Parcel 2 (35 acres), Parcel 3 (517 acres), Parcel 4 (112 acres), and Parcel 5 (430 acres). The Columbia Gateway river frontage spans Columbia River Miles (RM) 100-102 (Figure 16.1 on page 220). The port's intended use of this site was for planned development of water-related, heavy, and light industrial uses. Parcels 2 and 3 were designated for heavy industrial (MH) use and parcels 4 and 5 were designated light industrial (ML) by the City of Vancouver Zoning Ordinance; this fit into the port's growth plans for the site.

All environmental impact assessments conducted under the Washington State Environmental Act (and the federal National Environmental Act) begin with statements of purpose and need. These statements establish the rationale for development of a project or site. They also put the example environmental impact assessment into context with the location and regional setting.

13.3 Purpose

As part of its long-term planning process, the Port determined that development of Columbia Gateway would be necessary to meet its mission of providing economic benefit to the community through leadership, stewardship, and partnership in marine and industrial development.

The purpose of developing Columbia Gateway was to—

• Fulfill the port's mission of economic development.

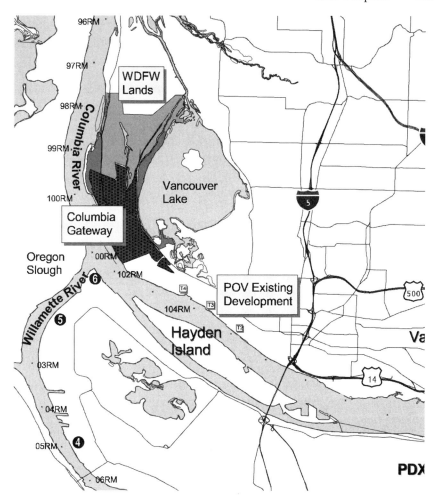

Fig. 13.1. Location of the Columbia Gateway area of the Port of Vancouver. The Columbia River is the border between the states of Oregon and Washington. This is a portion of one map included in the written EIS. (From [2]).

- Allow the port, the City of Vancouver, and Clark County to compete for economic development opportunities on a local, regional, national and international scale.
- Allow the port to continue to support diverse light-industrial, heavy-industrial, and water-related/water-dependent uses and multi-modal transportation facilities to meet and support future regional import and export needs.
- Provide development opportunities for businesses and services that support light-industrial, heavy-industrial, and water-related uses.
- Provide family-wage jobs and assist the city and county in meeting employment goals.
- Plan and develop Columbia Gateway (the last major industrial development lands on the Columbia River) while meeting or exceeding all environmental stewardship requirements and generating community consensus regarding Port development and key issues.
- Develop industrial lands in accordance with state, county, and city planning guidelines consistent with applicable local, state and federal regulations.

The first five purposes are objectives (or goals) of the project, while the last two purposes are constraints (or criteria). Therefore, the structure of the assessment must be such that decision-makers are presented with information suitable for a multi-objective/multi-criteria decision-making process.

13.4 Need

In January 1996, the State of Washington Superior Court for Clark County ordered the port to complete an EIS for Columbia Gateway Parcels 2–5 before development could proceed. To fulfill its mission and to comply with the court order, the port decided to develop a subarea plan and complete an EIS on that subarea plan. Columbia Gateway was a valuable economic resource for the following reasons—

- Columbia Gateway was the largest industrial property under one ownership in the Portland/Vancouver area.
- According to Clark County data, Columbia Gateway's Parcel 3 was the only available industrial area within the county with adequate

deep-draft river frontage to allow for future water-related industrial development.

- Clark County's Comprehensive Plan (1994) estimated a need for over 2,400 acres to accommodate 20-year county job growth forecasts and maintain a competitive industrial land supply. About three-fourths of the land would be needed for light-industrial use.

- Development of the Columbia Gateway was likely to play a key role in helping the county to expand its industrial base and provide family-wage jobs for a growing population. While it was difficult to predict the number of jobs Columbia Gateway development would provide, industrial lands most likely would be needed to accommodate them.

- The Port decided that one method of estimating the number of jobs that development of Columbia Gateway could bring was to use existing facilities as examples. In 2000, six hundred acres of the Port supported industrial use. Port businesses employed more than 2,170 people, and another 1,071 were employed off-site by operations directly relating to the port's industrial operations. From a different perspective, the port directly and indirectly provided about 5.4 jobs per acre. At 5 jobs per acre, the 1,094 acres at Columbia Gateway could have resulted in more than 5,400 jobs

- The Oregon Economic Development Department (OEDD) estimated that Clark County would capture approximately 24 percent of the regional industrial job growth by 2020. Therefore, industrial lands should be available to accommodate this growth. Given the shortage of industrial land in Clark County, future use of Columbia Gateway for purposes unrelated to industrial development would result in a loss of the potential for job creation and the county's ability to provide economic opportunities for a growing population. Expansion of Port operations in particular would be limited and would limit the Port's ability to adjust to changing markets. This would impact the public agency's financial health. If development occurred under the Subarea Plan, the agricultural and open space uses of the property would have been changed. Farming would have ceased and the natural resources on the property altered. If implementation of the plan did not occur, it was presumed that the present agricultural practices and open space nature of the property would continue.

14

Scoping

The scope of the written EIS for the Port of Vancouver (USA)'s Columbia Gateway master development plan was determined through broad participation. Discussions were held with City of Vancouver staff, port staff, and resource and regulatory agency representatives. The public responded to a scoping notice issued in July 1999, and additional comments were gathered at a scoping open house on August 19, 1999. The scoping notice was sent to approximately 500 citizens, agencies, and businesses. The scoping notice briefly described five alternatives developed as part of the master plan completed by the port in 1998. Alternative 1, a plan for the placement of dredge disposal material from the US Army Corp of Engineers Columbia River Federal Navigation Channel Improvement Project, originally included in the scoping notice, was later eliminated from consideration. Alternatives 2, 3, and 4 of the Subarea Plan could accommodate disposal of dredge material.

The scoping notice also included information about the project's history and location. The scoping notice was a request for affected agencies, Native American tribes, and members of the public to provide comments on any aspect of the proposed scope of the EIS. A summary of major comments relating to the scope of the EIS is presented below—

- The EIS should include a detailed discussion of potential adverse impacts associated with wetland and vegetation loss. It should also include a detailed discussion of the potential impacts to wildlife from permanent loss of riparian areas and agricultural lands.

- Analysis of environmental impacts to threatened and endangered species and natural resources should be science-based.
- The EIS should include analyses relating to water quality, fish predation, and critical habitat for the proposed boat basin and any in-water structures and the proposed pierhead line.
- The EIS should address the use of potentially contaminated sediments that may be used to fill Columbia Gateway to prepare it for development.
- Potential land use impacts from the Subarea Plan alternatives should be thoroughly addressed in the EIS.
- Potential impacts to cultural resources located at Columbia Gateway from Subarea Plan alternatives should be thoroughly addressed in the EIS.

During the scoping process, a number of agency representatives and members of the public expressed support for one alternative or another. Where possible, these comments were incorporated during the alternative refinement process. Comments collected during the scoping process also included suggestions for mitigating potential adverse impacts of alternatives.

14.1 Determining Components

The actual scoping process resulted in a decision to include these components in the environmental impact assessment:

1. Geography (Location)
2. Geology and Soils
3. Air quality
4. Noise
5. Wetlands
6. Hydrology and Water Quality
7. Vegetation and Wildlife
8. Fish
9. Environmental Health
10. Land Use
11. Light and Glare
12. Aesthetics
13. Recreation
14. Historic and Cultural Resources
15. Transportation
16. Jobs and Economic Growth
17. Utilities and Public Services Infrastructure

For this example of the modern approach to environmental impact assessments five of the above components will be used to illustrate the use of fuzzy sets and fuzzy logic in an approximate reasoning model. The five broad components represent environmental, social and economic concerns that have both quantitative and qualitative aspects to their values. The five components are:

1. Vegetation and Wildlife. These environmental components are closely related to each other and are almost certainly included in every environmental impact assessment. The two related aspects of terrestrial environments can be measured using any of many different standardized techniques and analyzed into useful information. However, unless there is very specific geographic isolation (as is the case with an island in a large ocean or an oasis in a desert), plant species composition will reflect secessional patterns controlled by climate and topography while wildlife populations reflect both resident and migratory species. The values associated with the measurements are an important consideration in determining significance.

2. Wetlands, Hydrology, and Water Quality. These are also related environmental components that have regulatory standards based on measurements as well as strong subjective aspects involved in assessing functions, values, and needs. Like other environmental concerns with numeric standards of acceptable condition, these three may be subject to the one-size-fits-all regulatory solution. This single number applied to all locations is much easier for assessing regulatory compliance but may have little relevance to what actually are the structure and processes at any specific location and project type. These three attributes (along with wildlife and fish) tend to generate the most controversy in the environmental impact assessment process.

3. Aesthetics. This is a purely subjective social component based on values and beliefs. It cannot be directly measured, but it has very strong emotional support among certain stakeholders. The cliché "beauty is in the eye of the beholder" describes the assessment of æsthetics. It is certainly difficult for people to define clearly and for assessment preparers to capture effectively.

4. Transportation. This is an economic component that also has measurable and subjective components. For a port-related development project's environmental impact assessment this category includes truck, rail and ship transportation. In this example only the

vehicular traffic is considered. It is straight forward to measure traffic density and travel time but the impact of changes on quality of life and other aspects is not so simple to calculate. Indirect aspects of transportation on land can overlap with air quality and public safety concerns. In this example, those peripheral factors will not be considered.

In the following example each of these components was separately considered, even expanded. Vegetation is a measure of total acreage of plant cover; threatened and endangered plant species present is a separate category. Wildlife is a measure of population size by species (or other taxonomic grouping); threatened and endangered populations is a separate category. Wetland size and wetland quality are two separate components in the example. Such refinement adds to the completeness of the assessment while requiring minimal additional effort.

14.2 Public Participation Process

Community members and agency representatives assisted in the development of the EIS. Two open house meetings were held. The first, on July 18, 2000, focused on refinements to the alternatives. The second, on April 25, 2001, focused on information generated by data analyses of current conditions. Open house attendees reviewed and commented on information related to wetlands, wildlife, vegetation, and cultural resources. A Technical Advisory Committee (TAC) also aided development of the EIS. The TAC was composed of representatives from organizations including the Port of Portland, the US Fish and Wildlife Service (USFWS), the Army Corps of Engineers, Washington Department of Fish and Wildlife (WDFW), Washington Department of Ecology (DOE), Washington Department of Transportation (WSDOT), Clark County Natural Resources Council, city and county staff, Greater Vancouver Chamber of Commerce, the Fruit Valley Neighborhood Association, and other stakeholders.

The purpose of the TAC was to assist the port and the project team in refining the alternatives, developing science-based analytic methods, reviewing technical information, and providing recommendations on issues related to the project at key points in EIS development. Smaller working groups made up of TAC members were established to allow better focus on specific topics and issues. The three working groups were Natural Resources, Land Use/Planning, and Infrastructure.

Three newsletters and two postcard updates were distributed during development of the EIS to keep community members informed. The newsletters included general project information, contact information, and updates on progress. The port provided updates through its web site and quarterly community newsletters, and its staff highlighted the project in its annual report, speakers bureau, and community tours. A mandatory, 30-day public review period followed release of the EIS to the public.

14.3 Conflicting Values

The stakeholders and other parties can be assigned to three groups: those who favored complete development of the Columbia Gateway's 1,094 acres to their maximum economic potential, those who favored no development at all and those who would be happy with some development between the extremes. The modern approach to resolving value differences is implemented by the following process.

The groups, or the individuals, are given a form with all the components listed by pairs and a copy of the table of values they are to use to express the relative importance of the two components to them, independent of any other beliefs or values (Table 11.1 on page 130). The collective values for each pair are averaged and that is the value used to determine the group consensus ranking of component importance.

For the purpose of illustrating how conflicting values can be quantified, each of the three stakeholder groups is assumed to have collectively compared all components pair-by-pair. The top eight components are shown in Table 14.1 on the following page.

There are several approaches to calculating results from the paired comparisons. A simple method uses the Octave[1] mathematical pro-

---

[1] GNU Octave is a high-level language, primarily intended for numerical computations. It provides a convenient command line interface for solving linear and nonlinear problems numerically, and for performing other numerical experiments using a language that is mostly compatible with Mat-Lab. It may also be used as a batch-oriented language. It is available at no cost.

Table 14.1. Results of pairwise comparisons of some components considered in the Columbia Gateway environmental impact assessment.

	Aesthetics	Transportation	H_2O Quality	Noise	Wetlands	Hydrology	Plant	Animal
Aesthetics	1.000	0.143	0.333	9.000	0.167	0.200	0.500	0.250
Transportation	7.000	1.000	0.200	9.000	0.333	0.143	1.000	4.000
H_2O Quality	3.000	5.000	1.000	5.000	0.250	0.167	0.125	3.000
Noise	0.111	0.111	0.200	1.000	0.143	0.125	0.500	0.200
Wetland	6.000	3.000	4.000	7.000	1.000	2.000	0.333	4.000
Hydrology	5.000	7.000	6.000	8.000	0.500	1.000	0.333	0.333
Plant	2.000	0.200	1.000	8.000	2.000	3.000	1.000	1.000
Animal	4.000	0.250	0.333	5.000	0.250	3.000	1.000	1.000

gramming language; MatLab® could also be used if this software is available. Each row of elements in Table 14.1 on the preceding page is written as comma-separated values in an ASCII-format file. In this example model the file is named *gateway.dat*.

These data are used in this Octave program, *ahp.m*:

```
# calculate primary vector
function m = ahp(x)
n = size(x,2);
r = prod(x,2).^(1/n); # .^ for element powers
m = r/sum(r);
```

The Octave program is run using the data file:

```
octave> data = load('gateway.dat');
octave> v = ahp(data);
octave> save -ascii gateway-scoping.output v
```

The output file has the principal vector extracted from this matrix with values listed in the same order as the rows in Table 14.1 on the facing page. The consensus importance ranking is shown in Table 14.2.

Table 14.2. Importance of components to be included in the example environmental impact assessment.

Component	Importance value
Wetlands	0.255
Hydrology	0.174
Vegetation	0.143
Transportation	0.134
Water Quality	0.130
Wildlife	0.101
Aesthetics	0.045
Noise	0.018

Among all three stakeholder groups the consensus is that wetlands accounts for a quarter of all concern about the site and æsthetics is of little concern, only 4.5 percent of the total. Noise was not considered in the example assessment because it was of almost no importance to the group participating in the process.

The views of all stakeholder groups are represented in the evaluation. Everyone's values have the same weight and are processed the same way. The result is a true consensus achieved without confrontation, delay, argument, or bruised feelings because everyone entered his values independently of everyone else.

The lead agency, or whomever is conducting the assessment, now has a technically sound and legally defensible list of priority concerns and can focus efforts on the components considered to be the most important. It might be decided, for example, that only components with a value greater than 0.10 (10 percent) will be considered in the assessment. The value of each of these components is different than would be calculated if all 19 components were shown in the example. Regardless, this ranking provides a rationale for expending time and effort on those that matter most to the stakeholders and other interested parties.

15

Baseline Environment

This chapter of the example presents the status of those components identified by the scoping process of the Columbia Gateway SEPA EIS as important to all stakeholders.

15.1 Vegetation and Wildlife

With the exception of Parcel 2, most of the current use of the Columbia Gateway site is agriculture: annual row crops and livestock grazing pastures. Wetlands were being created on Parcel 2 at the beginning of the last century. Upland areas and some wetlands on Parcels 3–5 are used for agriculture by four farms. Two farms have single-family dwellings on them. Overall, approximately 28 acres of the site are impervious surfaces created by buildings and paved roads, excluding Lower River Road, which divides Parcel 5. The jurisdictional shorelines of the site contain no significant development and are dominated by riparian vegetation.

Surrounding areas are used for recreation, wildlife habitat enhancement, and industrial purposes. Frenchman's Bar Park, a county recreational facility, is located along the shoreline of the Columbia River northwest of Parcel 5. The Shillapoo Wildlife Area, managed by the Washington State Department of Fish and Wildlife (WDFW), is north of Parcels 4 and 5. Vancouver Lake Park, east of Parcel 4, is a recreational facility frequented by fishermen, swimmers, and wind-surfers. Industrial facilities near the project area include Vanalco, a large aluminum-manufacturing plant located 0.25 miles (400 meters) south of Parcel 3, and a private recycling facility immediately south and west of Parcel 3.

15.1.1 Vegetation Cover Types

Vegetation categories are based on hydrology and dominant plant cover species. (Table 15.1). Seven upland vegetation types and four

Table 15.1. Areal extent (in acres) of each vegetation cover type identified on the Columbia

Gateway site.

Type	Parcel				
	2	3	4	5	Total
Riparian forest	2.5	30.4	00.	6.1	39.0
Forested wetland	0.0	1.1	0.0	1.1	2.2
Upland scrub-shrub	1.0	0.0	3.0	0.0	4.0
Scrub-shrub wetland	17.0	62.0	0.0	1.1	80.1
Upland pasture	14.5	142.1	4.5	257.9	419.0
Emergent wetland	0.0	190.2	76.5	134.8	401.5
Row crop	0.0	42.6	28.0	16.0	85.6
Seasonal slough	0.0	14.1	0.0	0.0	14.1
Developed land	0.0	9.3	0.0	14.0	23.3
Paved road	0.0	7.3	0.0	0.0	7.3
Sand	0.0	17.9	0.0	0.0	17.9
Totals	35.0	517.0	112.0	430.0	1094.0

wetland types are typical of river floodplains in the Pacific Northwest.

Row crop areas are nonwoody vegetation on hydric[1] and nonhydric soils. Farmed wetlands are included in the row crop category because of the similarity in vegetative structure. Annual row crops are grown on 386 acres within Parcels 3–5. The principal crops are corn, oats and barley. Alfalfa had been grown in earlier decades under different tenant farmers. A few areas on the site were not planted for the

---

[1] Wetland-type soils that are saturated within 30 centimeters of the surface for at least 2 weeks during the growing season.

year 2000 growing season. These areas contain a variety of forbs such as Canada thistle (*Cirsium arvense*), bull thistle (*Cirsium vulgare*), curly dock (*Rumex crispus*), false dandelion (*Hypochaeris radicata*), cudweed (*Gnaphalium* spp.), tansy ragwort (*Senecio jacobaea*), St. John's wort (*Hypericum perforatum*) and red clover (*Trifolium pratense*).

Upland pasture covers approximately 421 acres on Parcels 3–5. The dominant species in this cover type are herbaceous vegetation such as tall fescue (*Festuca arundinacea*), orchard grass (*Dactylis glomerata*), bentgrass (*Agrostis* spp.), and perennial ryegrass (*Lolium perenne*). The pastures are grazed by cattle and horses and are periodically mowed. Upland herbaceous areas on Parcel 2 are not grazed but are included in this category because of the similarity in vegetative structure. Herbaceous-dominated areas grazed by livestock but with wetland hydrology are categorized as emergent wetland.

The riparian forest category includes nonhydric areas that are dominated by woody plants taller than 20 feet (6.1 meters) and with canopy coverage greater than 50 percent. Mature, deciduous riparian forests are located along the Columbia River on Parcel 3 and along Buckmire Slough on Parcels 4 and 5. This vegetation community is spatially limited, totaling 38 acres. For the purpose of the habitat analysis, the portion of the riparian forest acreage adjacent to seasonal and permanent sloughs were included with the slough system acreage. The forested corridor on Parcel 3 is dominated by mature black cottonwood trees and has an understory of willow shrubs, cottonwood saplings, snowberry (*Symphoricarpos albus*), and Himalayan blackberry (*Rubus discolor*). The southeast portion of this riparian corridor is forested wetland. The patch of riparian forest in the northern tip of Parcel 5 is dominated by Oregon ash with an understory of snowberry. This stand of ash trees encompasses a seasonal slough dominated by reed canarygrass (*Phalaris arundinacea*). Himalayan blackberry is well established along the edges of this stand and in open areas within the stand. The riparian community located along Buckmire Slough between Parcels 4 and 5 is similar to the other riparian areas, but also contains bigleaf maple (*Acer macrophyllum*) and large, standing dead trees.

The forested wetland category is similar to the riparian forest category but contains wetland hydrology, plants, and hydric soils. Forested wetland on Parcel 3 consists of black cottonwood overstory and willow shrubs, cottonwood saplings, reed canarygrass, stinging nettle, willow herb (*Epilobium watsonii* or *E. ciliatum*), Himalayan blackberry, and nightshade (*Solanum dulcamara*) in the understory. The forested wetland on Parcel 5 is surrounded by upland forest composed of mature ash trees with an understory of willows, cottonwoods and reed canarygrass. Forested wetland is extremely limited on the property: approximately 1 acre on Parcel 5 and 1 acre on Parcel 3.

Upland scrub-shrub vegetation is dominated by woody plants less than 20 feet (6.1 meters) high with a canopy coverage between 30 and 50 percent. Scattered mature trees may be present, but shrubs are the dominant vegetation. Upland scrub-shrub is limited to 3 acres (1.2 hectares) on Parcel 4 and consists of tall willow shrubs and 20-foot (6.1 meter) black cottonwood trees. The herbaceous layer of this area is grazed by cattle and consists of nonnative grasses, Queen Anne's lace (also called wild carrot, *Daucus carota*), St. John's wort, false dandelion, curly dock, moth mullein (*Verbascum blatteria*), sweet vernal grass (*Anthoxanthum odoratum*), teasel (*Dipsacus sylvestris*), and Canada thistle.

Emergent wetlands cover 133 acres (53.9 hectares) on Parcels 3 and 5; the majority of this acreage is on Parcel 3. These wetlands are used primarily as pasture for livestock grazing. The wetlands are dominated by reed canarygrass with other emergent plant species including spike rush (*Eleocharis palustris*), velvet grass (*Holcus lanatus*), and bent grass. Emergent wetlands on Parcel 2 include two enhanced wetlands and a created wetland. These wetlands have been planted with a number of wetland plant species, including willow and cottonwood. Cattails (*Typha latifolia*) have colonized the wetlands on Parcel 2 and a scrub-shrub wetland on Parcel 3.

A dense patch of black cottonwood saplings and willow shrubs has established in a depressional wetlands on Parcel 3. The scrub-shrub wetland category is similar to the upland scrub-shrub habitat but with wetland characteristics. This vegetation category covers 15 acres with emergent vegetation such as water plantain (*Alisma plantago-aquatica*), reed canarygrass, spike rush, toad rush (*Juncus bufonius*), and cattail.

Seasonal sloughs include emergent wetlands, open water, and an upland fringe of shrubs and trees. The sloughs are remnants of old, truncated river meanders and are long, narrow, and mostly parallel with the Columbia River. Ponding occurs in the sloughs during winter precipitation and high groundwater. Open water is limited during summer because of evaporation, seasonal water table lowering and decrease in precipitation. Typical shrub species along the banks of the sloughs include willow and rose (*Rosa* spp.), with scattered black cottonwood and ash saplings. Himalayan blackberry is dominant upslope from the wet areas, while reed canarygrass is the dominant herbaceous species in the sloughs. A small patch of mature ash, Oregon white oak, and black cottonwood trees is adjacent to one of the remnant sloughs in Parcel 3.

15.1.2 Sensitive, Threatened, and Endangered Plant Species

Based on data from the Natural Heritage Information System managed by the Washington Department of Natural Resources (DNR), no rare plants or high-quality ecosystems occur in the vicinity of the project site. Prior studies, wetland delineations and reconnaissance field visits confirmed the lack of threatened, endangered, or sensitive species on the site. Listed and sensitive plant species that may have existed there, but are not currently present, include water howellia (*Howellia aquatilis*, federally threatened) and Columbia cress (*Rorippa columbiae*, species of concern). These two plant species have been recorded across the river from the project site at Sauvie Island, Oregon, and water howellia has been found on the Ridgefield National Wildlife Refuge, north of the property. Water howellia is an aquatic plant that once occurred throughout the Pacific Northwest. It has very specific habitat requirements and is susceptible to disturbance and slight changes in water levels. Potential suitable habitat for water howellia is absent on the property, because there are no undisturbed, natural wetlands. The survival of Columbia cress populations depends on unstable riparian substrates and periodic, catastrophic flooding. Damming and diking of the Columbia River system has eliminated much of the historic Columbia cress habitats. Potential on-site habitat is limited by the existing reed canarygrass and lack of major flood events.

15.1.3 Wildlife

The Columbia Gateway parcels have been disturbed by flooding (most recently in 1996), disposal of dredge materials, livestock grazing, and by farming practices. As a result, there no longer are unique or rare habitats. Surrounding areas such as Sauvie Island, Oregon, and areas around Vancouver Lake support similar plant and wildlife species. In a landscape context, the Columbia Gateway site serves as a link between

Vancouver Lake and the Columbia River as well as to the adjacent state wildlife reserve.

The site provides habitats, of varying quality and quantity, for a number of resident and migratory wildlife species. Descriptions of wildlife habitat conditions are based on the dominant plant cover types. Agricultural land, including row crops and pasture, is the most common habitat on the property, whereas forested wetlands is the most unique and limited habitat.

The agricultural areas and seasonal sloughs provide wintering and resting areas for thousands of migratory waterfowl. At least five different subspecies of Canada geese winter on the site, including cackling Canada geese (*B.c. minima*), lesser Canada geese (*B.c. parvipes*), Taverner's Canada geese (*B.c. taverneri*), dusky Canada geese (*B.c. occidentalis*), and Western Canada Geese (*B.c. canadensis*). The different subspecies form large mixed flocks on the site. The birds forage on grains and other vegetation from approximately October to April. WDFW limits hunting of the dusky Canada goose, a State Priority Species, because of its poor breeding success in Alaska. Sandhill cranes (*Grus canadensis*) use the site primarily as stopover habitat during migration where they forage in wet pasture and in cornfields. Dabbling ducks, including northern shovelers (*Anas clypeata*), Bufflehead (*Bucephala albeola*), northern pintails (*A. acuta*), and American wigeons (*A. americana*) have been observed foraging and resting in open water areas during winter. Wintering waterfowl habitat is abundant in the vicinity because of the Shillapoo Wildlife Area and Ridgefield National Wildlife Refuge as well as surrounding agricultural land located on Sauvie Island and other properties to the north. Adult and juvenile bald eagle (*Haliaeetus leucocephalus*) winter in the vicinity of the site partly because of the abundance of prey.

The seasonal sloughs and emergent wetlands have been disturbed by agricultural practices and cattle grazing but provide habitat functions for several species. These wet areas provide foraging habitat for great blue herons (*Ardea herodias*), egrets, and dabbling ducks and breeding habitat for amphibians such as Pacific treefrogs (*Hyla regilla*) and long-toed salamanders (*Ambystoma macrodactylum*). The long-toed salamander is a native species that is able to exploit a wide range of habitats, especially disturbed environments. Songbirds such as the red-winged blackbird (*Agelaius phoeniceus*) and the common yellowthroat (*Geothylpus trichas*) nest in the narrow band of vegetation that fringes the seasonal sloughs. Bullfrogs (*Rana catesbeiana*), an aggressive, non-native species, have been detected in some of the sloughs and reduce the quality of the habitat by preying on native reptiles, amphibians, and juvenile birds.

Small rodents such as deer mice (*Peromyscus* spp.), vagrant shrews (*Sorex vagrans*), Townsend's vole (*Microtus townsendii*), Townsend's mole (*Scapanus townsendii*), and pocket gophers (*Thomomys* spp.) also use the open agricultural lands on the site. The brush rabbit (*Sylvilagus bachmani*), opossum (*Didelphis virginiana*), Eastern cottontail (*Sylvilagus floridanus*), and raccoon (*Procyon lotor*) are also expected to occur in the pastures and along the edges of riparian habitat and seasonal sloughs. Pasture and row crops may not provide enough cover for larger mammals, but coyote (*Canis latrans*) and black-tailed deer (*Odocoileus hemoinus*) are likely to traverse along open area edges. A coyote den with five pups was found in 1997.

Another type of open habitat on the site is sandy beach on Parcel 3. This habitat is sparsely vegetated and does not support a large number of terrestrial species. Birds such as waterfowl or gulls may roost on the beach or nest near the pioneer plants established on the upper beach. Canada geese and goslings have been observed on the beach, indicating that a nest was nearby.

Because most mammals are secretive or nocturnal, the presence of mammals on the site was based on habitat conditions and signs such as tracks, scat or vegetation marks. The beaver (*Castor canadensis*) is expected to occur in the sloughs based on teeth marks in trees along the banks. River otters have been observed in the lowlands and are expected to use the sloughs during high water levels. Mink (*Mustela vison*) have been observed along the shores of sloughs. Muskrat are expected to occur on the site seasonally or in low numbers based on the lack of deep marshes and competition from the introduced Nutria (*Myocastor coypus*). Nutria consume large amounts of vegetation and are common in shallow waterways such as the on-site sloughs. A muskrat skull was found adjacent to a seasonal slough on Parcel 3 in 1997.

The largest, contiguous tract of habitat on the site, aside from the open, agricultural areas, is the riparian forest on Parcel 3. Other substantial areas of forested habitat are adjacent to the Buckmire Slough system along the northern boundary of Parcels 4 and 5. Most fragmented patches of habitat have been disturbed by cattle grazing and the disposal of dredge materials. The strip of riparian forest on Parcel

3 may have been left standing to serve as a shelter-belt for adjacent cropland.

The riparian forests and forested wetlands provide habitat for many wildlife species, including red-tailed hawks (*Buteo jamaicensis*), Cooper's hawks (*Accipiter cooperii*), black-capped chickadee (*Poecile atricapillus*), downy woodpecker (*Picoides pubescens*), yellow warbler (*Dendroica petechia*), Bullock's oriole (*Icterus bullockii*), American robin (*Turdus migratorious*), spotted towhee (*Pipilo maculatus*), cavity-nesting ducks and European starlings (introduced species—*Sturnus vulgaris*). All nest in, or along the edges of, the riparian areas. An active bald eagle nest is located in a large cottonwood tree adjacent to the riparian forest on Parcel 3. Adult long-toed salamanders are likely to occur in the riparian habitat under downed logs and woody debris. Treefrogs, which are also not strictly associated with wet areas, are also expected to use the on-site riparian forests. Common mammals that are expected to occur in riparian habitat include the raccoon, opossum, squirrels (*Sciurus* spp.), and Pacific jumping mouse (*Zapus trinotatus*). Mink are also associated with aquatic areas and are likely to occur on the site as long as prey is abundant.

15.1.4 Threatened and Endangered Wildlife Species

The presence of federal and state listed, proposed for listing, candidate species, and wildlife species of concern was determined by correspondence with federal and state wildlife agency staff and by reconnaissance-level field surveys. Bald eagle nesting and wintering activities occur on the property. The bald eagle, currently listed as threatened, is under review for removal from the federal list of threatened and endangered species. It would, however, still remain listed as threatened in the State of Washington even if it were federally delisted. Bald eagle nesting activities have occurred on the project site and in the immediate site vicinity since 1992). The current nest site on Parcel 3 (active since 1998) fledged two eagles in 1998, but failed to produce young during the 1999, 2000, and 2001 breeding seasons.

The Aleutian Canada goose (*Branta canadensis leucopareia*) is a state-threatened subspecies of the Canada goose that may occur in the Pacific Northwest during migration. The Aleutian Goose was removed from the federal endangered species list in February 2001. The Aleutian Canada goose is considered a coastal migratory species that does not winter in significant numbers in the project vicinity. Aleutian geese that may winter in the project vicinity are rare and are considered incidental stragglers. Additionally, only one or two individual Aleutians winter in the Willamette Valley (Oregon) region per 100,000 Canada Geese.

Sandhill cranes (*Grus canadensis*), a state endangered species, use the project site and surrounding farmland as stopover habitat during migration. On January 13, 2000, approximately 400 cranes were observed foraging in pasture grasses on Frenchman's Bar Park just north of Parcel 5 and approximately 10 cranes were observed loafing adjacent to an emergent wetland on Parcel 3). The sandhill cranes observed on-site and in the project area were not identified to subspecies. The cranes that occur in the Vancouver Lowlands include the lesser sandhill crane (*G. c. canadensis*), and the greater sandhill crane (*G. c. tabida*). The greater sandhill crane, the only subspecies that breeds in Washington, has nested at Conboy Lake Wildlife Refuge in Klickatat County (Washington) and at the Yakima Indian Reservation in Yakima County (Washington). The limited distribution, low breeding numbers and low fledgling success of greater sandhill cranes probably accounts for its listing; however, the state listing does not distinguish among subspecies.

No proposed or candidate terrestrial wildlife species are expected to occur on the Columbia Gateway site or in the vicinity. Terrestrial species of concern to federal resource agencies that may occur in the broad, general area include two bat species: the long-eared myotis (*Myotis evotis*) and the long-legged myotis (*Myotis volans*). Both bat species are found throughout the State of Washington and are associated with conifer forests. During the winter months they usually migrate to warmer climates. A bat survey was not conducted because potential suitable habitat (coniferous forest) is absent and the two bat species are not likely to be present. The Western toad (*Bufo boreas*) is also a species of concern at the federal level that is neither known to use the property nor is expected to be found here.

Priority Habitat and Species (PHS) mapping shows two Great Blue Heron rookeries located within 1 mile (1.6 kilometers) of the Columbia Gateway site. One rookery is located east of Parcel 3 and another rookery is located to the north of Parcel 5. Also, a heron rookery is located in Buckmire Slough on the northern tip of Parcel 4. PHS data also identifies wintering waterfowl habitat on Parcel 3, bald eagle nests on Parcels

3 and 5, cavity-nesting habitat for waterfowl along Buckmire Slough between Parcels 4 and 5, and perch trees used by eagles along Buckmire Slough.

15.2 Wetlands, Hydrology, and Water Quality

15.2.1 Wetlands

Wetland types at Columbia Gateway consist of four vegetative communities: forested, scrub-shrub, emergent and aquatic bed. The pattern of vegetative communities and plant species within them is summarized. Nineteen jurisdictional wetlands totaling 148 acres were identified on Columbia Gateway (Table 15.2). Emergent wetlands are the

Table 15.2. Wetland types and areal extent prior to development, in acres.

Wetland type	Parcel 2	Parcel 3	Parcel 4	Parcel 5	Total
Forested scrub-shrub	0	0	9.0	6.5	15.5
Forested	0	2.1	0	1.1	3.2
Scrub-shrub	17.8	21.8	0	0	39.6
Emergent	0	87.0	0	2.7	89.7
Totals	17.8	110.9	9.0	10.3	148.0

most prevalent (89.7 acres), whereas forested wetlands are a small component (3.2 acres). Functional site indices were calculated for each wetland based on its size and capability to perform a specific function (e.g., water quality, wildlife habitat). The mean functional site indices were relatively constant across all wetlands included in the study. Parcel 4 wetlands have a slightly higher general habitat suitability index than do those on other parcels because this parcel has several contiguous sloughs and greater connectivity among the different wetlands. Functions and values for wetlands on each parcel are summarized in Table 15.3 on the next page.

15.2.2 Hydrology

Columbia Gateway is within the 100-year floodplain of the Columbia River in an area called the Vancouver Lake Lowlands. The sequence

Table 15.3. Functions and values (0 = lowest, 10 = highest) for wetlands on each of the four parcels at Columbia Gateway.

Attribute	Parcel			
	2	3	4	5
Water quality improvement				
Sediment removal	10	10	10	10
Nutrient removal	6	5	6	6
Heavy metal and organics removal	4	4	5	4
Water quality control				
Peak flow reduction	10	10	10	10
Downstream erosion decrease	10	10	10	10
Groundwater recharge	4	4	5	4
Habitat functions				
General suitability	3	3	7	3
Invertebrate suitability	4	3	6	2
Amphibian suitability	2	2	6	2
Wetland-associated bird suitability	6	4	7	3
Wetland-associated mammals	2	4	9	3
Fish suitability	0	0	0	0
Native plant richness	2	1	3	2
Primary production and export	–	–	–	–

of land uses and flood that have influenced the wetlands and water quality are summarized here.

River discharge and tides influence water levels in the Columbia River at Columbia Gateway. River discharge is determined partially by precipitation and snowpack throughout the Columbia River drainage basin and partially by the volume of water released from behind Bonneville Dam. In the Gateway reach low water is at 1.6 feet (48.8 centimeters) NGVD[2]. The OHWM[3] is at 16.7 feet (5.1 meters) NGVD. The effect of the tide is greatest at low river stages when it may alter the water surface elevation by about 1.5 feet (45.8 centimeters). Effects are negligible at river stages deeper than 12 feet (3.7 meters) NGVD.

---

[2] National geodetic vertical datum
[3] Ordinary high water mark

The water surface elevation in the 100-year return frequency flood is estimated to be approximately 26.5 feet (8.1 meters) NGVD and would just overtop the berm on Parcel 3. The height of the berm averages 26 feet (7.9 meters).

A groundwater monitoring study was performed between 1997 and 1999 to examine the subsurface hydrology of Columbia Gateway in relation to wetlands and uplands. Changes in groundwater levels generally correspond to long-term changes in the Columbia River stage level, while short-term fluctuations in groundwater levels seem to be driven by precipitation rather than river stage. The elevation at which wetland hydrology is present varies from approximately 13 to 16 feet (4.0–4.9 meters) as a function of rainfall, river levels, and position within on-site drainage networks.

There is little or no runoff from Parcels 3 and 5. Precipitation either percolates directly into permeable surface soils or accumulates in ponds and wetlands where it evaporates or percolates into the ground. After entering the ground, it moves laterally, eventually discharging to the Columbia River. If 50 percent of precipitation evaporates and 50 percent percolates into the ground, the average daily discharge of groundwater to the Columbia River is 1.3 cubic feet per second. The calculation of daily discharge to groundwater is based on average annual precipitation (47.8 inches; 1.2 meters), infiltration rate (50 percent assumed) and parcel size (517 acres). The Clean Water Act requires regulation of post-development discharges if impacts (e.g., riverbank erosion) are anticipated.

15.2.3 Water Quality

Water quality in the lower Columbia River Basin has been degraded by human activities, including discharge of pollutants and alteration of the river's natural flow regime. The port constructed the flushing channel along the west side of Parcel 3 in the 1980s. The flushing channel's purpose is to improve water quality in Vancouver Lake by allowing water movement between the river and the lake diurnally and seasonally.

Both historical and current water temperature data from the lower Columbia River exceed the state's special temperature standard of 20°C in August.

Water quality characteristics in the lower Columbia River include—

- Water is well-oxygenated, consistently meeting the requirement that dissolved oxygen concentrations be at least 90 percent of saturation.
- Nutrient concentrations are low relative to those of most large US rivers.
- Concentrations of fecal bacteria are low and consistently below standards for water contact recreation (i.e., for a 30-day period average of 126 colonies per 100 ml water).
- Diazanon®, Eptam®, Metolachlor®, and Napropamide® were detected in some water samples taken near Hayden Island but none was at a concentration exceeding ambient standards established by DOE[4] and DEQ[5] pursuant to the federal Clean Water Act.

In addition to establishing ambient standards pursuant to the federal Clean Water Act, DOE and DEQ are required to list those river reaches in their states that do not comply with in-stream water quality standards after the application of conventional wastewater treatment technology to point sources of pollutants. These lists are known as 303(d) lists. When the DOE and DEQ 303(d) lists were last published in 1998, the lower Columbia River was in partial compliance with DOE and DEQ standards for ambient water quality. In the DEQ 303(d) list, the lower Columbia River was out of compliance for water temperature, total dissolved gases, arsenic, fecal coliform, and sediment bioassay.

Ambient water quality standards require that surface waters be within the range of pH 6.5 to 8.5. Natural surface waters usually lie within this range. Acidity or alkalinity of waters outside this range may adversely affect aquatic life. Two of 14 samples taken near Columbia Gateway exceeded the ambient standard. The pH was too high during April and May 1994, possibly as a result of increased algae production.

Total dissolved gas measures the amount of oxygen, nitrogen, and argon in water. Water oversaturated with gases can be harmful to fish. Of three measurements of total dissolved gas near Columbia Gateway in 1994, one exceeded the ambient standard of no more than 110 percent of saturation.

Median concentrations of dissolved metals in waters near the site are similar to those in other US rivers, and are in compliance with ambient standards for of aquatic life and human health. Arsenic, a human

---

[4] Washington State Department of Ecology.
[5] Oregon State Department of Environmental Quality.

carcinogen, was detected in all four water samples at levels that exceed ambient standards for the of human health.

The mass of pollutants discharged to the Columbia River in groundwater under Parcel 3 is very small. Parcel 3 is only partially used for agriculture; the rest is open land. The only substances likely to be deposited on the land surface are animal wastes and possibly insecticides and/or herbicides. Animal wastes could contribute nutrients and pathogenic organisms to the groundwater. Pathogenic organisms would be filtered out as a result of passage through the soil and are unlikely to reach the river. Insecticides and/or herbicides may have accumulated in low concentrations in the soil and may reach the river.

15.3 Aesthetics

The written report logically begins with a definition of æsthetics from Webster's dictionary:"the beauty of art and/or nature." The term applies to Columbia Gateway as the site's visual appeal to the observer. With regard to a vacant, yet developable, site such as Columbia Gateway, æsthetics can be considered to be the appeal of the site in its various stages, from vacant and undeveloped to fully developed.

In the professions of land-use planning and development, æsthetic factors generally are considered to include site design and layout, building architecture, construction materials, colors and textures, and the use of landscaping to enhance or complement structures.

Because the site and its immediate vicinity lack extensive development, little tangible evidence of developed (or the "built environment") æsthetic values exists. However, the property's currently undeveloped nature may have æsthetic value for those who value the preservation of open space regardless of condition or use.

The farms on the property include typical farm structures: dwelling, barn, and related outbuildings. In terms of age and architectural style, none of these structures is historic, unique, or unusual; they are typical of local agricultural buildings erected during the latter half of the twentieth century.

Most of the 1,094 acres are vacant open space used for grazing or row crops. There are dispersed stands or clusters of vegetation throughout, including several varieties of mature trees, dry land and wetland vegetation, and scrub-brush.

Two public parks, Frenchman's Bar Riverfront Park and Vancouver Lake Regional Park, are north and east (respectively) of Columbia Gateway. South of the site are the facilities for Tidewater Barge Co., Russell Towboat and Moorage Company, Columbia Resource Company recycling plant, Vanalco aluminum plant, Bonneville Power Administration (BPA) power substation, and Columbia Public Utility's River Road Generating Plant. These industrial facilities are separately housed in structures unique to each individual function and land use. The tallest structure in the area, aside from the BPA transmission towers, is the River Road Generating Plant emissions stack (approximately 198 feet [60.4 meters] tall).

The Columbia River is the western border of the property and is largely inaccessible and partially hidden from view by existing vegetation. The nearest public access roads are SR[6] 501 and Old Lower River Road. The Columbia River is not visible from either of these roads within the Columbia Gateway area. SR 501 is one mile from the Columbia River, while old Lower River Road is approximately three-quarters of a mile from the river. Overall, the Columbia Gateway view is limited to farm operations, industrial operations to the south, and the two regional parks. The parks are not very visible because of the lack of height of any part of the park development and the way in which the park facilities blend into the local landscape. Distant views of the site are from the northeast, across Vancouver Lake. Overall, the lack of urban development on Columbia Gateway results in little identifiable or definable æsthetic value with respect to the built environment.

15.4 Transportation

15.4.1 Introduction

The written EIS describes transportation conditions in both 2000 and 2020 in the vicinity of Columbia Gateway. These conditions are presented as a comparison of potential trips generated by each of the alternatives.In addition, the City's concurrency ordinance will require traffic analyses to be submitted with development applications as the parcels are developed. Transportation improvements may be required to meet concurrency ordinance requirements.

---

[6] State route.

Existing transportation conditions include vehicular operations, public transportation, nonvehicular transportation facilities and planned improvements. The transportation network analyzed in the written impact statement extends eastward from Columbia Gateway to Interstate 5 (I-5) along the Fourth Plain, Mill Plain, and Fruit Valley corridors, and includes existing public streets within all Port property.

Interstate-5 is the region's major north-south limited-access road corridor for commuters and freight traffic. It carries average daily traffic (ADT) of approximately 120,000 vehicles per day across the Interstate Bridge[7], decreasing to 80,000 vehicles north of the 39th Street interchange in Washington. From the north, Fourth Plain and the Fourth Plain/I-5 interchange provide access to Columbia Gateway. Mill Plain and the Mill Plain/I-5 interchange provide access from the south.

15.4.2 Existing Traffic Volumes and Operations

Intersections typically are the points of greatest delay and congestion in urban transportation systems due to the number of potential conflicting movements that require assignment of right-of-way using traffic signals, stop signs, and other traffic control devices. Staff from WS-DOT[8] and the City of Vancouver selected eight intersections in the study area for analysis during afternoon peak-use conditions. These locations represent known and potential locations of congestion and/or traffic operations concerns. Turn movement counts conducted in June 2000 were adjusted to reflect anticipated conditions after the opening of the Mill Plain Extension. Traffic operations analysis was conducted using methods in the 2000 Highway Capacity Manual, which assigns a level of service (LOS) to intersection operations ranging from A (free-flowing conditions) to F (operational breakdown). Table 15.4 on the facing page presents intersection LOS criteria.

Table 15.5 on page 212 summarizes the afternoon peak hour LOS in 2000 with the Mill Plain Extension in place, based on the adjusted 2000 intersection volumes and geometry. The table also shows intersection volume-to-lane capacity (v/c) ratios, another measure of intersection performance. The v/c ratio is the ratio of conflicting traffic turning

---

[7] Spanning the Columbia River and connecting Portland, Oregon ,with Vancouver, Washington.

[8] Washington State Department of Transportation.

Table 15.4. Level of service categories for intersections with and without traffic control signal lights. For intersections with signals the numbers are the average seconds each vehicle is stopped. For unsignaled intersections the numbers are the average seconds of total delay for each vehicle.

Level of service	Signaled	Unsignaled
A	< 10	< 10
B	> 10, < 20	> 10, < 15
C	> 20, < 35	>15, < 25
D	> 35, < 55	>25, < 35
E	> 55, < 80	< 35, < 50
F	> 80	> 50

movements to the theoretical intersection capacity, which is based on characteristics such as intersection lane geometry, traffic control, and traffic composition. As the table indicates, all study area intersections are functioning at LOS D or better with existing traffic. Based on afternoon peak period traffic counts conducted for the analysis, truck traffic ranges from 10 percent to 15 percent of total intersection traffic at the western end of the study area, decreasing to less than five percent on the eastern end. Existing levels of truck traffic were held constant for the analysis of future conditions.

15.4.3 Nonautomobile Transportation Facilities

This section describes existing nonautomobile/truck transportation facilities serving Columbia Gateway, including public transit service provided by C-TRAN, public-use aviation facilities, and existing rail, bicycle and pedestrian facilities.

15.4.3.1 Public Bus Transportation

C-TRAN operates one bus route through Columbia Gateway. Route 1 (Fruit Valley) runs from the 7th Street Transit Center to the Frito-Lay manufacturing/distribution plant via Mill Plain and Fourth Plain. It provides service every half-hour, operating 6:00 AM to 9:00 PM weekdays, 7:45 AM to 8:00 PM Saturdays, and 8:45 AM to 6:00 PM Sundays. Route 1 carries about 500 passengers daily.

Table 15.5. Existing traffic conditions at certain signaled and unsignaled intersections in the vicinity of the Columbia Gateway area (data from afternoon peak hour surveys in 2000; delay time in seconds).

Location	LOS	Delay	V/C
Signaled intersections			
Fourth Plain Blvd. and Mill Plain Blvd.	**B**	18.0	0.25
Fourth Plain Blvd. and Fruit Valley Rd.	D	45.8	0.63
Fourth Plain Blvd. and Main St.	D	41.6	0.78
Mill Plain Blvd. and Columbia St.	B	17.9	0.73
Mill Plain Blvd. and I-5 South Ramps	C	27.1	0.87
Mill Plain Blvd. and I-5 North Ramps	C	22.2	0.77
Unsignaled intersections			
NW Lower River Road and Port entry			
Major approach: westbound	A	9.4	
Minor approach:	C	16.4	
Fourth Plain Blvd. and W. 26 St. extension			
Major approach: westbound	A	9.6	
Minor approach: northbound 26 St.	C	17.9	

15.4.3.2 Airports

Portland International Airport (PDX) provides regional and international passenger and freight air service. It is located approximately 10 miles southeast of Columbia Gateway. Pearson Airpark is a small, public-access, general aviation airfield located approximately five miles east of of Columbia Gateway. Pearson Airpark serves small aircraft used mostly for recreational and business charter travel.

15.4.3.3 Bicycle Facilities

There are bicycle lanes on Lower River Road (SR 501) from Fruit Valley Road to the west, on Mill Plain Boulevard, and on Fruit Valley Road between Fourth Plain and 39th Street.

15.4.3.4 Pedestrian Facilities

Fourth Plain Boulevard, Mill Plain Boulevard, Columbia Street and Main Street have sidewalks on both sides. West of downtown, there are sidewalks on Mill Plain Boulevard, Fruit Valley Road, and the north side of Fourth Plain Boulevard. NW Lower River Road and NW Old Lower River Road have paved shoulders rather than sidewalks.

15.5 Fuzzifying Initial Conditions

For each of the components included in the environmental impact assessment a linguistic variable is created for the appropriate range of values. For example, the *Vegetation* component (Figure 15.1) might represent the riparian, wetland, and other such cover types typical of an undisturbed environment adjacent to a major river. The universe of discourse is from 0 acres to 1,000 acres (close enough to the project site size of 1,094 acres) and there are five qualitative terms describing the amount of desired vegetation: *Tiny*, *Small*, *Moderate*, *Large*, and *Huge*.

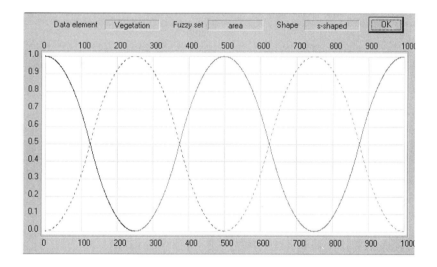

Fig. 15.1. The amount of vegetation of value to wildlife under existing conditions.

Some components are either complex aggregates of factors (water quality, for example) or concepts that have no underlying measurement (æsthetics, for example). The universe of discourse for such linguistic variables can be defined as convenient. There may be situations when the Water_quality linguistic variable needs to be assigned within the interval [0,10] rather than the interval [0,1.0]. Water_quality has two fuzzy terms, *Unacceptable* and *Acceptable* (Figure 15.2). Other com-

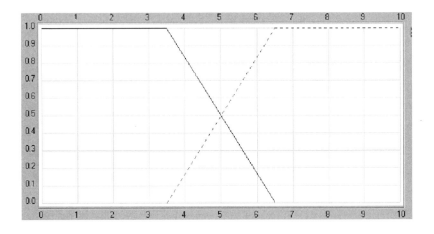

Fig. 15.2. The fuzzy term sets *Unacceptable* and *Acceptable* in the linguistic variable "Water_quality."

ponents in the assessment are also represented by linguistic variables and fuzzy term sets. In a production model used for a particular environmental impact assessment, there will be a linguistic variable for each component in order to understand the baseline conditions and changes with each alternative. These linguistic variables are not used directly in the calculation of the Environmental Condition Index (ECI). Instead the values are translated to grades of membership in alternative fuzzy sets representing the linguistic variables *Good* and *Not_Good* for each component (Figure 15.3 on page 218).

For the purposes of illustration with the Port of Vancouver's Columbia Gateway development assessment, seven components are used in ten linguistic variables:

1. **Vegetation** represents the total area classified as wetlands, sloughs, and riparian forest, as these types are of greater value than are the uplands. The linguistic variable is partitioned into five fuzzy term sets: *Tiny*, *Small*, *Moderate*, *Large*, and *Huge*. The universe of discourse is the open interval (0–1,000) acres.

2. **TE_plants** is the linguistic variable representing the percentage of total plant species found on the site that are listed as rare, threatened, endangered, or otherwise of special concern. The universe of discourse is the open interval (0,100). The fuzzy term sets are *Very_few*, *Few*, *Some*, *Many*, and *Predominant*.

3. **Wildlife** represents the number of wildlife species identified by surveys of the site. Not considered in the example are the different types (e.g., waterfowl, raptors, small mammals, large mammals, amphibians), and the intensity or duration of use (migratory stopover, breeding, foraging). The universe of discourse is the open interval (0–1,000) individuals. The fuzzy term sets are *Low*, *Moderate*, and *High*.

4. **TE_fauna** is the wildlife equivalent of TE_plants. It has the same universe of discourse and fuzzy term sets.

5. **Wetland_size** captures the amount of jurisdictional[9] wetlands on the site. The universe of discourse is approximately the site size, 0–1,000 acres. The fuzzy term sets are *Tiny*, *Small*, *Moderate*, *Large*, and *Huge*.

6. **Wetland_quality** is a generalization of values and functions as defined by the appropriate local authorities. The universe of discourse is an arbitrary scale in the open interval (0,1000) and the fuzzy term sets are *Low*, *Moderate*, and *High*.

7. **Hydrology**, in the example, represents the percentage change in runoff to the Columbia River by different amounts of development in the alternatives. The universe of discourse is the open interval (0,100) and the fuzzy term sets are *Very_small*, *Slight*, *Moderate*, *Large*, and *Heavy*. In a production model Hydrology would be computed by IF-THEN rules with antecedents including slope, vegetation cover, soil moisture, and precipitation.

8. **Water quality** is an amalgam of dissolved oxygen, pH, toxins, temperature, and other parameters of interest. The universe of discourse is in the open interval (0,10) with two fuzzy term sets, *Unacceptable* and *Acceptable*. In a production model where water quality is an important consideration it would be a submodel in which in-

---

[9] Those wetlands that meet criteria requiring regulatory approval for removal or fill. Mitigation is always a requirement for permission to alter the wetland.

teractions among low values and threshold-exceeding values are evaluated in detail.

9. **Aesthetics** in the example has an arbitrary scale from 0 to 10 as the universe of discourse and three fuzzy term sets: *Low, Normal,* and *High.*

10. **Traffic** represents transportation in the example. It is the relative change in Level of Service for road traffic at the designated intersections. The universe of discourse is the open interval (0,100) with three fuzzy term sets: *Small, Moderate, Large.*

In this example the TE_plants and TE_fauna linguistic variables are not included in the calculation of the baseline Environmental Condition Index because no species is listed as threatened or endangered under federal or state criteria. The symmetric summation inference method for combining fuzzy sets (Section 9.10.4 on page 104) accommodates missing data, and that is what those two variables are under baseline conditions.

Using available input data for the baseline environmental conditions, the fuzzified grades of membership are—

Vegetation: $\mu_{moderate} = [0.96]$
Wildlife: $\mu_{low} = [1.00]$
Wetland_size: $\mu_{small} = [0.67]$
Wetland_quality: $\mu_{low} = [1.00]$
Hydrology: $\mu_{slight} = [0.72]$
Water_quality: $\mu_{acceptable} = [1.00]$
Aesthetics: $\mu_{nice} = [0.75]$
Traffic: $\mu_{moderate} = [1.00]$

15.6 Environmental Condition Index

After baseline measurements have been fuzzified using membership functions of the appropriate linguistic variables, the membership grades need to be interpreted and combined in a meaningful way. The result of these computations is an environmental condition index in the interval [0.0,1.0]. The two reasons membership grades are not directly used for such an index are that each component of the baseline conditions has a different importance weight and the membership grade has no intrinsic meaning with regard to the condition of the assessment site.

Importance weights are value judgments expressed by stakeholders who participate in the scoping process. Calculated using a no-cost, readily available software package as shown in Chapter 14 on page 187, these weights are also presented in Table 15.6[10] and they will

Table 15.6. Importance weights of each environmental component included in the example assessment.

Component	Weight
Wetlands	0.255
Hydrology	0.174
Vegetation	0.143
Traffic	0.134
Water Quality	0.130
Wildlife	0.101
Aesthetics	0.045

be used in the calculation of the environmental condition index.

Finding meaning in the membership grades of the assessment components is a two-step process. The first step determines the "goodness" of the measured value of each environmental, social and economic component of the assessment site. As pointed out earlier, the concept of "goodness" is a linguistic concept that is not directly measurable; it is a reflection of individual values and beliefs. Of equal importance, it is not valid to mix together values representing size, rate, length or other different measures; they need to be translated to a common measurement scale. The second step calculates the weighted symmetric summation [32] of the ratio of "goodness" to "not-goodness" as the environmental condition index.

Goodness is represented by a fuzzy membership function, *Good*, that increases from $\mu_{Good} = [0.0]$ to $\mu_{Good} = [1.0]$ in an S-shaped curve. The universe of discourse is the range from 0 to 100 (Figure 15.3). As *Good* increases its complementary fuzzy set, *Not_Good* decreases from $\mu_{NotGood} = [1.0]$ to $\mu_{NotGood} = [0.0]$. For each assessment component the degree of *Good* is determined by applying a block of rules that

---

[10] Noise is not included in the example because of insufficient data and low importance value (1.8 percent) to stakeholders.

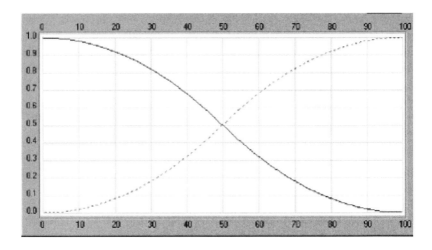

Fig. 15.3. The fuzzy sets for *Good* and *Not_Good* used to quantify environmental conditions at the Columbia Gateway site.

translates the component's grade of membership in a fuzzy set to a degree of goodness.

For the existing conditions, the ECI is calculated to be 0.80. This, then, is the standard against which all project alternatives will be compared. Of equal importance, the ECI for the existing conditions represents the quantification of all measured values, observations, and relative importances of various environmental, social and economic components that are normally described as words, table and figures. This quantification removes the onus from the decision-maker in understanding what the text, tables, and figures mean.

Project Alternatives

The Port of Vancouver (USA)'s Columbia Gateway written environmental impact assessment has four alternatives. These alternatives were originally developed as part of a master plan completed by the port in 1998. The four master plan alternatives considered in the environmental impact assessment were refined as part of the EIS development process in order to define the range of uses, characteristics, and intensity of development that could occur at Columbia Gateway. The alternatives are presented in the written EIS in terms of proposed use(s) on each of the individual parcels into which the total area has been divided. These parcels are presented in Figure 16.1 on the next page. A fifth alternative—full build-out of the property—has been added to this example. Whereas the three development alternatives in the Master Plan included on-site mitigation, the fifth alternative includes extensive off-site mitigation, particularly in the state-owned wildlife area immediately to the north of the Gateway area.

The five alternatives that make up the Columbia Gateway draft subarea plan are:

1. **No Action**. Under this alternative, no development occurs. Existing farm and agricultural uses are assumed to continue until the leases expire. When and if farming operations on Parcels 3 and 5 cease, the land would return to unused fallow ground.

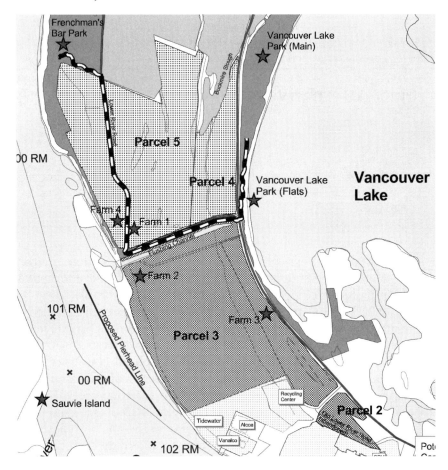

Fig. 16.1. The four parcels (numbered 2–5) that compose the Columbia Gateway area at the Port of Vancouver (USA). Parcel 1 is the developed area, part of which is seen at the bottom center of the map. (From [2]).

2. **Parcel 3 Water Development.** This alternative develops 504 acres of Parcel 3, which would include 47 acres of water-dependent uses[1] located within the first 200 feet landward of the ordinary high wa-

---

[1] Water-dependent uses are those intended primarily for commercial, public, and recreational uses that require direct contact with the water and cannot exist at a nonwater location due to the intrinsic nature of the operation.

ter mark (OHWM) of the Columbia River and Vancouver Lake Flushing Channel. Water-dependent uses include those which cannot logically exist in any location but on the water. Examples include, but are not limited to, water-borne commerce; terminal and transfer facilities; ferry terminals; watercraft sales in conjunction with other water-dependent uses; watercraft construction, repair, and maintenance; moorage and launching facilities; aquaculture; log booming; and public fishing piers and parks. The remainder of Parcel 3 (457 acres) is proposed for water-related development; those industrial and commercial activities that are related to water-dependent ones but do not require direct access to water. Examples of water-related activities are hotels, restaurants and other stores serving crews and passengers as well as port workers. Mitigation would include 242 acres on Parcels 4 and 5. Total development proposed under this alternative is 504 acres.

3. **Parcel 3 Heavy and Water Development/Parcel 5 Light-Industrial Development.** In this scenario, 420 acres of Parcel 3 (including 20 acres of water-dependent uses located within the first 200 feet landward of the OHWM of the Columbia River, 132 acres of water-related uses located between 200 feet and 1,000 feet landward of the river and 268 acres of heavy industrial development beyond 1,000 feet landward of the river) would be available for development. This alternative also proposes 280 acres of light industrial development and a 20-acre public boat basin on Parcel 5. Mitigation would include 326 acres (84 acres on Parcel 3, 112 acres on Parcel 4, and 130 acres on Parcel 5). Total development proposed under this alternative is 720 acres.

4. **Parcel 3 Water Development/Parcel 5 Light-Industrial Development.** Development on 420 acres of Parcel 3, which would include 20 acres of water-dependent uses located within the first 200 feet landward of the OHWM of the Columbia River. The remainder of development on Parcel 3 would include 400 acres of water-related uses. This alternative also proposes 280 acres of light industrial development and a 20-acre public boat basin on Parcel 5. Mitigation would include 326 acres (84 acres on Parcel 3, 112 acres on Parcel 4, and 130 acres on Parcel 5). Total development proposed under this alternative is 720 acres.

5. **Marine and Terminal/Heavy-Industrial, Light-Industrial, and Commercial Development.** This alternative was not in the Environ-

mental Impact Statement but has been added for completeness in discussion and analysis. Under this alternative, all 1,094 acres will be developed into marine shipping facilities (breakwaters, docks); terminal facilities; and other water-dependent, water-related, and nonspecific heavy- and light-industrial operations. Commercial operations would be located on areas adjacent to the main road, Vancouver Lake Flushing Channel, and state wildlife reserve on the north end. Such commercial development would feature "grass and glass" business parks. Mitigation, as appropriate, would be integrated in the development areas but most mitigation would be off-site in the wildlife reserve.

16.1 The Affected Environment

16.1.1 Vegetation and Wildlife

Potential impacts to vegetation and wildlife habitat from each development alternative were evaluated in the written EIS using the Habitat Evaluation Procedures (HEP), a method endorsed by the project's Technical Advisory Committee, federal resource agencies, and most wildlife research programs. HEP is used to determine the habitat quality of a site based on the suitability for selected animal species of dominant vegetation cover types. Impacts were calculated based on the footprint of development under the assumption that full implementation of a development alternative would occur in one year. Species selected to represent the dominant cover types on the site were great blue heron, mallard duck, wintering Canada goose, savannah sparrow, pond breeding amphibians, mink, black-capped chickadee, and yellow warbler.

In the written EIS, habitat suitability is determined using mathematical models that produce a Habitat Suitability Index (HSI) for each species and each cover type. Not all cover types provide habitat for all of the evaluation species. HEP quantifies habitat in terms of habitat units (HUs), which are calculated by multiplying the number of acres of a habitat for a particular evaluation species by the HSI. One HU is equivalent to one acre of the best habitat for a species (with an HSI of 1), and could also be represented by two acres with an HSI of 0.5, or four acres with an HSI of 0.25.

The HEP is a method for quantifying the baseline value of the area for wildlife groups and provides a mechanism to judge how much value might be lost under a particular development scenario. Mitigation provides wildlife value to compensate for impacts. The HEP also was used to calculate the area of a specific cover type to compensate for losses. The usefulness of habitat is measured by the Habitat Suitability Index, so replacement for wildlife habitat does not have to be of exactly the same.

16.1.1.1 Alternative 1

This alternative would result in no impacts to the existing baseline habitat conditions other than those produced by normal environmental variability and agricultural use of land. If no impervious structures are created, the existing pasture habitat is expected to undergo ecological succession from bare soil to upland grass habitat to scrub-shrub and eventually to riparian forest.

16.1.1.2 Alternative 2

This alternative results in the loss of 857.4 HUs on Parcel 3, including 309.5 HUs of pasture, 158.7 HUs of emergent wetland, 142.6 HUs of row crop, 130.6 HUs of seasonal slough, 75 HUs of riparian forest, 38.2 HUs of scrub-shrub wetland, and 2.8 HUs of forested wetland. The area of greatest impact would be to row crops, although the greatest habitat impact would be to upland pasture (309.5 HUs). The next greatest habitat impact would be to emergent wetland, resulting in the loss of 158.7 HUs.

16.1.1.3 Alternatives 3 and 4

Both alternatives would result in the loss of 1,151.9 HUs on Parcels 3 and 5, comprising 652.7 HUs of upland pasture, 186 HUs of row crop, 127.4 HUs of emergent wetland, 74.1 HUs of riparian forest, 70.5 HUs of seasonal slough, 38.2 HUs of scrub-shrub wetland, and 2.8 HUs of forested wetland. Alternatives 3 and 4 would result in the greatest amount of impact to habitats and wildlife species at Columbia Gateway. The cover types most impacted by these alternatives include upland pasture and row crops.

16.1.1.4 Alternative 5

This alternative involves building out the entire 1,094-acre property. On the north end, adjacent to the wildlife area, commercial development of business parks ("grass and glass") would provide landscaped grounds, but not deliberately designed wildlife habitats. Almost all of the area would become impervious surface for heavy and light industry, shipping facilities, maritime trade support, transportation interchange and employee/visitor parking. Functionally, all habitat units (HUs) are lost.

16.1.2 Threatened and Endangered Species

16.1.2.1 Bald Eagle

According to the US Fish and Wildlife Service (FWS), bald eagle breeding and wintering activities occur almost all year. Nesting can occur from January 1 to August 15, and winter activities occur between October 31 and March 31 of the following year. Alternatives 2–5 are evaluated with the assumption that construction of infrastructure for marine industrial development within 200 feet of the OHWM (the extent of the shoreline jurisdiction) on Parcel 3 will require the removal of some mature black cottonwood trees, including the bald eagle nest tree, landward of the OHWM.

Under Alternative 1, there would be no identified, deliberate, short-term impact on existing habitat conditions. In the long term, bald eagle habitat would be sustained or could decrease depending on the future use of Port property now in private agricultural leases. The other alternatives would involve removing the tree used by the eagles as a nest site. However, the birds had used a different nest on the property in addition to the current one, and there are many more suitable trees adjacent to the river in the wildlife area north of the Port's property. Removal of the nest tree when unoccupied, and the adults off elsewhere, would be no more traumatic to the birds than if a former nest tree was lost from wind throw or fire.

16.1.2.2 Sandhill Crane

The sandhill crane is a state threatened species that uses Columbia Gateway only as a stopover during migration. Impacts to sandhill crane habitat are evaluated qualitatively based on a review of a sand-hill crane management plan. Habitats considered as potential for foraging by sandhill cranes include row crop areas, upland pasture, and emergent wetland. Because the annual row crops are fully harvested in the fall, such areas are not considered high quality feeding areas for migrating sandhill cranes. Therefore, all impacts are minimal and much better, more consistent habitats are available in the wildlife reserve next door.

16.1.3 Wetlands, Hydrology, and Water Quality

16.1.3.1 Wetlands

Wetlands would be affected directly and indirectly by the development of Columbia Gateway. The most obvious direct impact is the grading and filling of wetlands. Direct impacts to wetlands were measured for each proposed alternative. The calculations assumed full build-out of the alternative scenarios. Potential impacts to wetlands are summarized in Table 16.1. Wetlands were categorized using the City of Vancouver's regulations.

Table 16.1. Direct wetland impacts for each alternative (in acres) (The natural processes that would change wetlands were not estimated in the written EIS).

| Category | Alternative | | | | |
	1	2	3	4	5
Forested	–	0.8	0.8	0.8	3.2
Scrub-shrub	–	21.8	22.7	22.8	39.6
Emergent	–	87.0	58.3	58.3	89.7
Forested scrub-shrub	–	1.3	1.3	1.3	3.2
Total	–	110.9	83.2	83.2	148.0

16.1.3.1.1 Alternative 1

Under this alternative, no development or filling would occur at Columbia Gateway. There would be no short-term adverse direct or indirect impacts on wetlands as a result of development. Current agricultural practices would continue until the leases were no longer available to the private landowners. Over time, there could be an improvement in the site's groundwater recharge, water quality, and habitat value. However, many of the wetlands would fill in with sediments and dead plant materials as is normal with plant succession. At least one wetland on the property was created as mitigation for prior development. However, it was not set below the water table. If it is not watered during the summer it dries completely. Other wetlands on the property will almost certainly follow the same path toward uplands.

16.1.3.1.2 Alternative 2

This alternative would impact the second largest acreage of Parcel 3 wetlands. All would be filled for a total impact of 111 acres. However, because no development is proposed on Parcels 4 and 5 under this alternative, the existing wetlands on Parcels 4 and 5 will remain as they are, with normal changes affecting them as would occur under alternative 1, the "No Action" alternative.

Construction adjacent to wetlands could cause clearing of understory and dumping of debris into wetland and buffer areas. Other potential effects are the introduction of invasive, nonnative plant and animal species that establish quickly in disturbed areas.

Without incursion to the existing wetlands in parcels not scheduled for construction, the addition of impermeable surface and changed soil characteristics can result in minor fluctuations in groundwater and inundation frequency, depth and duration. These changes can result in changes in plant and animal species composition. Altering hydrology, water quality, plant composition, or introducing invasive nonnative species can reduce the presence of native species and limit the desired functions of the wetlands.

In this alternative, 112 acres of Parcel 4 and 130 acres of Parcel 5 would be used as mitigation. Mitigation would likely consist of both enhancement of existing wetlands and creation of new wetlands. The long-term effects of the mitigation are expected to be positive, improving water quality and hydrologic connection after the detailed construction plans and specifications are developed for the wetland mitigation site.

16.1.3.1.3 Alternative 3

Partial development of Parcel 3 and light industrial development on Parcel 5 is proposed under this alternative and 83 acres of wetlands would be filled. On Parcel 3, eighty-two acres would be filled. On Parcel 5, less than one acre of palustrine scrub shrub wetland would be included in the filled area. This alternative would impact less wetland on Parcels 3 and 5 than would Alternative 2. However, the wetlands remaining intact on Parcel 3 and 5 are expected to be indirectly impacted development activities.

In Alternative 3, a total of 112 acres of Parcel 4 and 130 acres of Parcel 5 would be used as mitigation. Mitigation would likely consist of both enhancement of existing wetlands and creation of new wetlands. The long-term effects of the mitigation are expected to be positive, improving water quality and hydrologic connection after the detailed construction plans and specifications are developed for the wetland mitigation site.

16.1.3.1.4 Alternative 4

Partial development of Parcel 3 and light-industrial development on Parcel 5 are proposed under this alternative. Fill would be placed on 420 acres of Parcel 3 and 280 acres of Parcel 5 at a depth sufficient to raise the sites, excluding the flushing channel and mitigation areas, above the 100-year floodplain elevation.

The amount of wetland to be filled on Parcel 3 under Alternative 4 totals 82 acres. The amount of wetland to be filled on Parcel 5 totals less than 1 acre of palustrine scrub-shrub wetland. Alternative 4 would impact less total area of wetland than Alternative 2 on Parcels 3 and 5. However, the wetlands remaining intact on Parcel 3 and 5 are likely to be indirectly impacted by development on Parcels 3 and 5.

In Alternative 4, a total of 112 acres of Parcel 4 and 130 acres of Parcel 5 would be used as mitigation. Mitigation would likely consist of both enhancement of existing wetlands and creation of new wetlands. The long-term effects of the mitigation are expected to be positive, improving water quality and hydrologic connection after the detailed construction plans and specifications are developed for the wetland mitigation site.

16.1.3.1.5 *Alternative 5*

All surface area would be fully developed under this alternative. However, storm water detention and treatment facilities would be constructed wetlands but the rest of the acreage would be filled above the 100-year floodplain elevation. Mitigation would be in the state wildlife refuge to the north which suffers from very poor habitat values.

16.1.3.2 Hydrology

Hydrologic change is not subject to direct regulation but may be regulated indirectly. If development at a site significantly increases peak runoff flow, the banks of the receiving stream may be destabilized (depending upon their composition, slope and ground cover). Sediment from eroding banks may decrease water quality. Because of this, some NPDES[2] stormwater permits require that post-development peak flows must not exceed pre-development peak flows. There is no such requirement in the port's NPDES permit because the receiving water—the Columbia River—is too large to be adversely affected by hydrologic discharge at the port. This lack of effect is documented in the alternatives evaluations. For each alternative, incremental increase in discharge to the river is compared to average daily flow.

All analyses indicate that change in runoff from the proposed development would have negligible effect on flow in the river. Change in available flood storage caused by development projects would be very small relative to total flood storage in the entire basin. Consequently, alterations in water surface elevations during the 100-year flood event would be negligible, as would the effects on upstream and downstream properties.

Potential significant water quality and hydrologic impacts of the proposed alternatives were separated into two categories: short-term impacts that occur during construction and long-term impacts that occur during operations.

16.1.3.2.1 *Alternative 1*

The "No Action" alternative proposes no development or filling at Columbia Gateway, so there would be no direct or indirect impacts on

---

[2] National Pollution Discharge Elimination System, the permitting system for point discharges under the Clean Water Act.

water quantity. Over time, there could be an improvement in the site s groundwater recharge. Alternatively, agricultural impacts could increase erosion into the receiving water. However, no deliberate development will cause change to the hydrologic conditions of the property.

16.1.3.2.2 *Alternative 2*

On Parcel 3 fill would be placed on 504 of the 517 acres at sufficient depth to raise the entire site above the 100-year floodplain elevation. Construction of ship berthing facilities under this alternative would require dredging approximately 250,000 cubic yards of material and maintenance dredging of 50,000 cubic yards at four-year intervals. This activity would occur outside shallow-water habitat area. The dredging would not have measurable hydraulic effects on the river and would not affect hydrology of the uplands, except for groundwater recharge rates.

Nearly the entire area of Parcel 3 would be occupied by buildings and paved surfaces. Stormwater runoff would be collected in catchbasins and routed to treatment ponds. Stormwater runoff and any spillage from offshore vessel loading facilities would be collected and routed to the treatment ponds. After treatment, stormwater would be discharged to the Columbia River.

Virtually all precipitation would become surface runoff and would be discharged to the Columbia River within a few hours. The estimated total annual volume of runoff would be approximately 1,750 acre-feet per year. This is equivalent to an average daily flow of 2.5 cubic feet per second. The incremental increase in discharge to the Columbia River would be 1.2 cubic feet per second. Average daily flow in the Columbia River adjacent to the port is greater than 200,000 cubic feet per second. The increase in flow would have a negligible effect on the hydraulics of the Columbia River.

16.1.3.2.3 *Alternative 3*

Fill would be placed on 420 acres of Parcel 3 and 280 acres of Parcel 5 to raise the sites, excluding the Flushing Channel and mitigation areas, above the 100-year floodplain elevation. Impacts on water quality, hydrology, and hydraulics are equivalent to those noted above but approximately 50 percent greater because of the increased affected land area.

Stormwater treatment under this development alternative would be the same as above. Almost all precipitation would become surface runoff and would be discharged to the river within a few hours. The estimated total annual volume of runoff from Alternative 3 would be 2,360 acre-feet. This is equivalent to an average daily flow of 3.4 cubic feet per second. The incremental increase in discharge to the Columbia River as a result of Alternative 3 would be 1.6 cubic feet per second. At peak flows, the increase would be less than 0.0008 percent, which would be a meaningless change to river hydraulics.

16.1.3.2.4 Alternative 4

Dredging and construction period impacts on hydrology would be the same as those described for Alternative 2, but scaled larger. It is expected that 420 acres of Parcel 3 and 280 acres of Parcel 5 would be occupied by buildings and paved surfaces. Stormwater runoff would be collected in catch-basins and routed to treatment ponds. Stormwater runoff and any spillage from offshore vessel loading facilities would be collected and routed to the treatment ponds. After treatment, stormwater would be discharged to the Columbia River. The estimated total annual volume of runoff from this alternative would be 2,360 acre-feet; discharge to the Columbia River would increase by an average of 1.6 cubic feet per second. This increase in discharge would have a negligible effect on the hydraulics of the river.

16.1.3.2.5 Alternative 5

While this alternative is predicated upon total site build-out, the design is still constrained by regulation to ensure that there is no measurable effects on river hydraulics. The other development alternatives have such small anticipated increases to river flow that extrapolation to complete build-out also results in low, single-digit increases to river water volume and discharge.

16.1.3.3 Water Quality

Soils at the proposed site are permeable,[3] so most precipitation occurring during construction would be expected to percolate into the

---

[3] There is no cemented substratum or clay lens under the surface.

ground. Any fuel or chemicals in use during construction would have to be properly stored or covered to avoid leakage into the groundwater or exposure to rainfall that would wash contaminants into the soils.

Pollutant loads characteristic of representative urban runoff within industrial lands, and the respective pollutant removal rates for treatment ponds, are shown in Table 16.2. Post-development level projec-

Table 16.2. Representative water quality of stormwater runoff from industrial lands in urban environments. Treatment efficiency based on removal within detention ponds.

Pollutant	Concentration (mg/l)	Treatment efficiency (%)
Total suspended solids (TSS)	194	70
Total copper (Cu)	0.053	50
Total zinc (Zn)	0.629	50
Dissolved copper (Cu)	0.009	0
Total phosphorus (P)	0.633	30
Total petroleum hydrocarbons (TPH)	2.792	50

tions for each of Alternatives 2–4 are shown in Table 16.3 on the following page. Alternative 1 was assumed to be no different from current conditions and Alternative 5 was not explicitly analyzed. The assumption for the effects of Alternative 5 would be approximately 48 percent greater than that for Alternatives 3 and 4.

16.1.3.3.1 Alternative 1

Under the "No Action" alternative, no development or filling would occur at Columbia Gateway, so there would be no direct or indirect impacts on water quality. Therefore, over time, there could be an improvement in the site's water quality, or agricultural use could result in

Table 16.3. Estimated annual discharge of pollutants post-development (in pounds).

Pollutant	Alternative			
	2	3	4	5
Total suspended solids (TSS)	291,725	394,986	394,986	584,579
Total copper (Cu)	133	180	180	266
Total zinc (Zn)	1,576	2,134	2,134	3,158
Dissolved copper (Cu)	45	61	61	90
Total phosphorus (P)	2,221	3,007	3,007	4,450
Total petroleum hydrocarbons (TPH)	6,997	9,474	9,474	14,022

greater runoff and decreased water quality in the river. This alternative was not projected into the future for potential changes.

16.1.3.3.2 Alternative 2

Industrial wastewater and sanitary sewage from development within the areas described in Alternative 2 would be discharged to the City of Vancouver's wastewater system for collection, treatment and disposal system. After treatment, industrial and sanitary wastewater would be discharged to the Columbia River in accordance with the provisions of the City's NPDES permit. No adverse effects on water quality are expected to occur.

In general, pollutants accumulate on paved surfaces and wash into the storm drain system with stormwater runoff. Annual pollutant loads were calculated using typical pollutant emission rates for industrial land in the Portland, Oregon, metropolitan area. Some proportion of the pollutant load would be removed in the treatment ponds. No violations of ambient water quality standards as a result of the construction of Alternative 2 is anticipated.

16.1.3.3.3 Alternatives 3 and 4

As with Alternative 2, industrial wastewater and sanitary sewage from development would be discharged to the city's wastewater collection, treatment, and disposal system and then to the Columbia River. No adverse effects on water quality is anticipated.

Although development under this alternative would increase the amount of pollutants discharged to the Columbia River from the site, it would not be expected to have a significant adverse effect on water quality because the discharge conditions are regulated in the same City of Vancouver NPDES permit as in Alternative 2.

As indicated in Table 16.3 on the preceding page, the estimated pollutant quantities from stormwater runoff under this alternative are the same as those estimated for development on Alternative 3.

16.1.3.3.4 *Alternative 5*

This alternative assumes full build-out and the maximum areas of impermeable surfaces of any alternative. The quantities processed by the treatment facilities and discharged to the river are approximately 48 percent greater than under Alternatives 3 and 4. However, these are still within the permitted annualized quantities.

16.1.4 Aesthetics

The original impact statement used three chapters from the municipal code of the City of Vancouver as impact assessment standards. The three chapters are Title 20 (Zoning Ordinance), Title 21 (SEPA Ordinance), and the Shoreline Management Master Program (SMMP). Compliance with these ordinances contributes to establishing æsthetic values for a proposed development. Title 20 addresses standards for light-industrial and heavy-industrial development, including building height, setbacks, lot coverage by buildings, landscaping, parking, signs, and outdoor storage. Title 21 permits evaluation of æsthetics in any proposed development. The SMMP includes development standards for projects within the shoreline area.

16.1.4.1 Alternative 1

The "No Action" alternative provides no opportunities for improvements to existing æsthetic values. However, it also contributes no additional impacts on existing views and vistas, so there is no need for measures to reduce adverse impacts.

16.1.4.2 Alternative 2

Under the development scenario of this alternative, æsthetics would be altered by structure height. Whether the new vistas are an improvement or an undesired outcome depends on beliefs and values. The views of Columbia Gateway from Frenchman's Bar Riverfront Park, Vancouver Lake Regional Park, the residential areas on the north and east sides of Vancouver Lake, the Columbia River, and from the opposite shore would be altered with the perceived change affected by distance from the site. However, the proposed industrial development is not inconsistent with existing industrial development to the southeast, and the entire property is zoned for heavy and light industrial activities as well as commercial use.

Use of certain building materials will provide mitigating values, the potential height of the structures cannot be mitigated by use of building materials, colors, shapes, and other design features. Views from all directions would be reduced; this component of æsthetic values of development is considered by some to be a negative impact. However, the city's 10-foot setback requirement along primary or secondary arterials, such as SR 501, will help to minimize the effect of new industrial development.

16.1.4.3 Alternative 3

The primary difference between this alternative and the previous one is the mitigation areas proposed along SR 501 and the Flushing Channel. The potential types of industrial development would be similar to Alternative 2. Therefore, relative to æsthetics, Parcel 3 development would be different only in maintaining mitigation acres along SR 501 and the Flushing Channel undeveloped as natural open spaces. These mitigation areas would provide a break in the continuous nature of industrial development. For drivers along SR 501 and visitors to the nearby public parks, mitigation may be perceived as an improved æsthetic environment because development will be less visible. Parcel 5 is zoned for light industrial development and height is limited to 45 feet (generally four stories above grade) unless there are increased setbacks in exchange for increased heights. Standards for the light industrial zoning district generally require less site coverage and enclosure/screening of processes and outdoor storage. The heavy industrial zoning district calls for greater setbacks but does not require screening or enclosure. However, the city's 10-foot setback requirement along primary or secondary arterials, such as SR 501, and landscape requirements required for light industrial development, will help to offset views of new industrial development.

Development of the 20-acre boat basin on the southwesterly corner of Parcel 5 will have little impact on the perception of æsthetic values, because the boat basin will not be visible from most surrounding areas. However, there may be some æsthetic impacts to boaters on the river and to people who view the area from the opposite side of the river. This is a value judgment and will be considered æsthetically neutral.

16.1.4.4 Alternative 4

This alternative is expected to provide the same æsthetic values before and after development as the previous one. No difference exists between the types of uses that may occur in Alternatives 3 and 4 related to the æsthetic impacts these uses might create. Therefore, the impacts to æsthetic values created by development under Alternative 4 would be the same as the impacts created by development under Alternative 3.

16.1.4.5 Alternative 5

Under this scenario, the entire 1,049 acres would eventually be developed to the maximum extent permitted under the three City of Vancouver code Titles noted at the beginning of this section. Within the bounds of the site development of structures and supporting facilities would certainly alter the views and perceived æsthetics of the property. However, this is the only alternative that applies mitigation to off-site areas, particularly the wildlife refuge immediately to the north. Because the vegetation in this area represents benign neglect rather than directed plantings to maximize habitat and human æsthetic values there is greater potential under this alternative than with any of the other four to increase regional æsthetics while the Columbia Gateway site itself is developed for its best potential use of industry and commerce.

16.1.5 Transportation

With this component it is easier to project changes under the "No Action" alternative because traffic volumes and densities increase in growing urban areas over time. The City of Vancouver (along with other municipalities in southwest Washington) had conducted a study to predict traffic during the afternoon peak hour in the year 2020. The written environmental assessment used that study as a future baseline. Predictions for the other alternatives are even more uncertain because they are so dependent on the mix of heavy industry, light industry, commercial development, and maritime trade. Rail and sea traffic will increase along with road traffic but only the latter was considered in the alternatives analysis of the written document.

Alternatives 2–4 were assessed in the written report to identify a worst-case alternative for more detailed analysis. The alternatives were screened using potential trip generation as the primary selection criterion. Trip generation serves as an overall indicator of other transportation measures and impacts, such as roadway travel delay and congestion, peak hour intersection levels of service, roadway improvement needs and costs, potential environmental and wetlands impacts from roadway widening, and impacts on (and improvement needs for) non-motorized travel modes such as walking and bicycling.

For the transportation analysis, it was necessary to develop more detailed assumptions for specific land uses. Table 16.5 compares daily and afternoon peak-hour trip generation for each alternative, including estimated truck trip generation. Alternative 1, or "No Action," would generate no new daily traffic from the Columbia Gateway property. Alternative 2, which includes no development in the 430-acre Parcel 5, is anticipated to create 3,125 daily trips. Of these trips, 545 are expected to occur within the afternoon peak hour. The general land use designations for Alternative 3 could result in some 12,075 trips per day, including about 1,910 during the afternoon peak hour. Trip generation from Alternative 4 would be only slightly less intense, with the potential for 1,670 trips during the afternoon peak hour, and about 11,435 on a typical weekday. Auto terminal trip generation was estimated assuming a ship in port, which represents worst-case conditions. Truck traffic was estimated separately for the land uses in each alternative, based on available sources of truck trip generation rates. Table separates daily and peak hour truck and auto traffic for each alternative.

Peak hour truck traffic would be similar in absolute terms with any of the build alternatives, in the range of 150 to 200 truck trips during the peak hour.

16.1.5.1 Alternative 1

Intersection turning movements for the afternoon peak hour with no development in Columbia Gateway area (the "No Action" alternative) were estimated using base year (1994) and 2020 traffic volume projections from the regional travel demand forecasting model maintained by Southwest Washington Regional Transportation Council (SWRTC). These projected peak hour "No Action" intersection turning movements are presented in Table 16.4 and can be compared with the current conditions presented in Table 15.5 on page 212.

As shown by the levels of service for the unsignalized intersections on Lower River Road in Table 16.4, traffic growth in the vicinity of Columbia Gateway would be negligible under the 2020 "No Action" scenario. Most of the study area intersections analyzed would operate at LOS D or better during the afternoon peak traffic hour. However, intersection failure is projected under future conditions at the intersection of Fourth Plain Boulevard and Main Street, compared to LOS D operations today. This change is due to anticipated traffic growth in the downtown area.

16.1.5.2 Alternative 2

Transportation analysis for this alternative was based on a high-traffic-generating, yet realistic, mix of specific land uses on Parcel 3. It was assumed that development of Parcel 3 would include an auto terminal, a high-volume bulk terminal, a liquid bulk terminal, and water-related industrial development. Under this scenario, approximately 3,125 daily trips could be generated, including 545 trips during the afternoon peak hour. Parcels 4 and 5 would not be developed under Alternative 2.

16.1.5.3 Alternative 3

As Was done with Alternative 2, an example of the mix of development that could occur on Parcel 3 was generated. In Alternative 3, develop-

Table 16.4. Level of service (LOS) and volume:capacity (v/c) ratios or intersections in the Columbia Gateway vicinity as projected for peak afternoon hour in 2020. Delay time in seconds.

Location	LOS	Delay	V/C
Signaled intersections			
Fourth Plain Blvd. and Mill Plain Blvd.	C	22.5	0.35
Fourth Plain Blvd. and Fruit Valley Rd.	D	46.4	0.66
Fourth Plain Blvd. and Main St.	F	81.5	0.82
Mill Plain Blvd. and Columbia St.	C	28.2	1.00
Mill Plain Blvd. and I-5 South Ramps	C	32.3	0.96
Mill Plain Blvd. and I-5 North Ramps	C	25.1	0.77
Unsignaled intersections			
NW Lower River Road and Port entry			
Major approach: westbound	A	9.6	
Minor approach:	C	18.6	
Fourth Plain Blvd. and W. 26 St. extension			
Major approach: westbound	A	9.6	
Minor approach: northbound 26 St.	C	17.3	

ment assumed the inclusion of water-related industrial and heavy industrial development. Also, 168 acres of light industrial development (on 280 acres) was assumed for Parcel 5. This acreage reflects the 60 percent maximum allowable ratio of building coverage to land area under City development regulations. Approximately 12,075 daily vehicle trips could be generated, including 1,910 trips during the afternoon peak hour. The 20-acre boat basin was not considered; rather, to ensure a conservative transportation analysis, the 20-acre boat basin was considered to be light industrial development, which generally generates more traffic than do recreational facilities.

16.1.5.4 Alternative 4

Alternative 4 is similar to Alternative 3 with water-related uses replacing heavy industrial uses proposed for Parcel 3. For the trip generation comparison, water-related industrial use under Alternative 4 was assumed to include an auto terminal, a liquid bulk terminal, and a heavy bulk terminal. This alternative could generate approximately 11,435 daily vehicle trips, including 1,670 trips during the afternoon peak hour. As in Alternative 3, the 20-acre boat basin was not considered separately but included as light industrial development.

16.1.5.5 Alternative 5

While this complete built-out alternative was not evaluated in the written assessment, the numbers are estimated as 48 percent greater than those for Alternative 4.

As shown in Table 16.5, potential daily and peak hour trip generation is greatest with the land use designations in Alternative 5; second greatest traffic volumes are projected to occur in Alternative 3. Potential trip generation serves as a representative value for overall traffic and transportation impacts. In other words, the alternative generating the most traffic is the alternative expected to generate the worst traffic impacts.

Table 16.5. Potential trip generation for each alternative.

Alternative	Daily			Afternoon peak		
	Trucks	Autos	Total	Trucks	Autos	Total
1	0	0	0	0	0	0
2	740	2,385	3,125	150	395	545
3	1,135	10,940	12,075	180	1,730	1,910
4	1,275	12,160	11,205	205	1,465	1,670
5	1,887	17,848	19,735	303	2,168	2,471

16.1.6 Cumulative Effects

The original Columbia Gateway EIS defined cumulative impacts as the range of additive or synergistic effects of the alternatives of a proposed project combined with other actions likely to occur at the same time. There are several cumulative benefits of the Columbia Gateway and other planned projects. These benefits include the potential for more efficient use of Port terminal property, increase in the local employment base and improvement in the local economy from both an increase in deep-draft ship traffic and the attraction of additional marine industry into the lower Columbia River.

If all the projects in the region, including the Columbia Gateway development, were approved and fully implemented, the regional environment would be changed. Some of the anticipated cumulative effects include—

Geography, Geology, and Soils: Development may result in groundwater levels' rising to near the surface during flood events and/or extended, intense rainfall.

Air Quality: The potential exists for a gradual degradation of air quality, depending on the sequence of development, type of industry, and amount of vehicular and vessel traffic.

Noise: The potential exists for a gradual increase in noise due to increased use of industrial equipment in the Columbia Gateway area. The intensity and duration of the noise would depend on the type, the implementation schedule (louder during construction activities), and total magnitude of industrial development.

Wetlands, Hydrology, and Water Quality: Cumulative analysis was focused on the types of dredging activities that could occur at one time within this portion of the lower Columbia River. This focus was reasonable because in-water work such as dredging has the potential to be disruptive and negatively impact water quality. The water quality could be temporarily impaired because of increased turbidity as sediment is resuspended.

Dredging is permitted only when potential impacts to anadromous fish within this section of the Columbia River are minimal. The potential for disturbance is routinely evaluated during the permit application stage.

Other possible perturbations to the river system are from stormwater runoff, minor runoff from upland construction during high precipitation events, or other inadvertent direct influences to water. The filling of low areas in the Columbia Gateway parcels is necessary to provide developed areas above the water table. When this fill is coupled with other past, present, and possible future fills there may be a cumulative net decrease in floodwater storage in the Columbia River floodplain. However, most of the flood storage losses along the lower Columbia River have already occurred. A few additional losses can be expected as ports are expanded. Losses could be offset by floodplain and wetland restoration projects, several of which are proposed for the region. Consequently, the cumulative impacts of the proposed project and other similar projects are expected to be modest in scale.

Vegetation and Wildlife: There could be cumulative loss of existing open space and associated wildlife habitat with development at Columbia Gateway and other project sites. However, proposed compensatory mitigation which is planned to be of much greater function and higher values for food and other animal needs. Therefore, if the mitigation (particularly that designated for the adjacent wildlife refuge) is carefully implemented there would be no significant adverse impacts to vegetation and wildlife from development at Columbia Gateway.

Fish and Aquatic Habitats: There is nothing proposed by this or any other known project that would affect adult fish, their migratory corridors or off-channel spawning areas. Development of berthing facilities involved extensive modification to the river bank and near-bank river bottom. However, once construction is completed, there are few areas of direct habitat or food resource losses for juvenile fishes. For anadromous[4] fish, there is the potential for disruption of migratory movement and feeding if all such habitats are depleted along many miles of river bank. However, by design, the north bank of the Columbia River (i.e., the side in Washington State) has stretches of developable land interspaced with stretches of protected lands (such as the state wildlife refuge adjacent to the Port).

Environmental Health: An increase in activities associated with the transport or handling of hazardous materials might increase the potential for environmental health risks from spills or other accidents. Compliance with existing environmental regulations could be expected to minimize such risks.

Land Use: Initial improvements at the Columbia Gateway site (e.g., placement of fill and infrastructure) would increase the amount of industrial land immediately available for development, while reducing the amount of land currently in open space or agricultural-related uses. Full build-out of Columbia Gateway and other projects

---

[4] Fish that spawn and rear in freshwater but spend the majority of their life in the ocean.

would result in a reduction in the inventory of industrial lands available for development in the Urban Growth Area (UGA). In turn, this may result in a need to expand the City's urban growth area to provide a 20-year supply of industrial lands.

Light and Glare: Although impacts of glare can be readily mitigated through the use of illumination shields and screening, additional industrial facilities would create new sources of light. The cumulative effects of such light on migratory birds and other animals was not evaluated.

Aesthetics: Existing open space at Columbia Gateway and other development sites would be replaced by buildings and other transportation facilities such railroad yards, intermodal transfer facilities, surface parking and other components of the human-built environment. This does not have the æsthetic appeal of open spaces, cows, and row crops. Whether this change is considered positive or negative is related to whether one has a job created by the development or if one is from some other place without a vested economic interest in the site.

Recreation: Activities at Columbia Gateway and other industrial sites in the vicinity could diminish the recreational experience at nearby facilities as a result of noise, light, and traffic. At the same time, development projects in the area could enhance recreation by providing new trail corridors, boating facilities and other recreational opportunities.

Historic and Cultural Preservation: As development occurs in the vicinity of Columbia Gateway, cultural resources are likely to be encountered and/or disturbed if any are present. However, from the perspective of cumulative impacts, the concern is minimal. All workers are regularly instructed to look for artifacts during site preparation. The local tribes and historical societies also have quite comprehensive inventories of land use of interest and value to them. Therefore, the potential for negative cumulative effects is negligible.

Transportation: Substantial increases in vehicular, rail and vessel traffic could result from development in the Columbia Gateway vicinity. These increases would occur over a long time. Compared to current operations, vehicular intersection failure[5] is projected to

---

[5] That is, traffic volume and accident increases.

occur with the "No Action" alternative at Fourth Plain Road and Main Street. This change is from anticipated traffic growth in the downtown Vancouver area. No cumulative effects analysis was done for transportation scenarios under the other alternatives.

Navigation: Conflicts between commercial and recreational water-borne traffic could increase as a result of the proposed recreational boat basin and the construction of new docks for maritime trade at both Columbia Gateway and other projects in the vicinity. No other cumulative impact to navigation from the Columbia Gateway project is anticipated with respect to an improved navigation channel combined with the new access and berthing area associated with the proposed project. When estimating cumulative impacts of marine terminal development on navigation, the overall trend in ship size and efficiency must be considered. Slow growth in vessel calls, despite new deep-draft facilities at Kalama, St. Helens, Vancouver, and Portland, is expected because the trend to larger vessels with increased carrying capacity per voyage results in fewer vessels in port at any period of time. New vessel calls generated by new marine terminals are offset by industry trends to increased carrying capacity per vessel. The planned deepening of the Columbia River navigation channel would allow even larger vessels onto the system, contributing to greater carrying capacity with fewer vessel trips.

Utilities and Public Services: Services are limited by existing infrastructure to individual businesses (such as inadequate rail, road, and parking facilities as well as the limited capacities of storm drains, electrical distribution systems and water supplies). Additional development would require new infrastructure. As projects are approved, utility master plans should be updated to provide an accurate picture of demand and supply. The anticipated cumulative effects are neutral; that is, more utilities and public services will be required, but they will be provided because this is a natural consequence of development.

17

Results and Conclusions

17.1 Running the Model

For purposes of illustrating how a fuzzy system model can be set up to calculate membership functions and condition indices, some Alternative 1 data and program code for the FuzzyEI-Assessortexttrademark is shown. The input data are described in the previous chapters. The specifics will depend upon the software used in a given assessment.

First, linguistic variables and fuzzy term sets are defined—

```
: vegetation -- wetland, riparian and other
:   higher-quality types
declare vegetation
  acreage flt
  area fzset (Tiny Small Moderate Large Huge);
: Vegetation area from 0-1,000 acres.
memfunct vegetation area s-shape
  Tiny     -1E6   0    0 250
  Small      0  250 250 500
  Moderate 250 500 500 750
  Large    500 750 750 1000
  Huge         750 1000 1000 +1E6;
: Wetland quality
declare wetland_quality
  wetqual flt
  quality fzset (Low Moderate High);
```

```
memfunct wetland_quality quality linear
  Low          -1E6   50 150 250
  Moderate     150 250 450 550
  High         450 550 700 1E6;
: hydrology -- increase in runoff over baseline
:    conditions, as percentage
declare hydrology
  hydro flt
  runoff fzset (Very_small Slight Moderate
                Large Heavy);
memfunct hydrology runoff normal
  Very_small -1E6   0   0 25
  Slight           0 25 25 50
  Moderate        25 50 50 75
  Large           50 75 75 100
  Heavy           75 100 100   +1E6;
: aesthetics -- varies with individuals.
declare aesthetics
  beau flt
  beauty fzset (Ugly Nice Beautiful);
memfunct aesthetics beauty s-shape
  Ugly          -1E6 0.0 1.0 4.0
  Nice           1.0 5.0 5.0 9.0
  Beautiful      6.0 9.0 10.0 1E6;
```

The above, while model specific, shows variable declarations, fuzzy sets within each linguistic variable, the universe of discourse, and the support set for each fuzzy term. The vegetation coverage, a simplified measure for the example, is based on a single measured quantity. Wetland quality is a constructed variable that summarizes a set of physical and chemical measurements of functions and values. The hydrology component is a measure of change over baseline conditions Aesthetics is shown as a Type-1 fuzzy set that represents a numeric expression of attractiveness. The model can be much more sophisticated than the one shown here since the incremental increase in computing time for complexity is minimal.

Once the membership functions for the linguistic variables are defined, the applicable rules must be provided. Here are shown only rules for fuzzifying the input values—

```
:rule r0 rule block 0
: (goal Fuzzify acreage in Vegetation area)
IF (in vegetation acreage = <acreage>)
  THEN fuzzify 1 area <acreage>;
:rule r4 rule block 0
: (goal Fuzzify wetqual in Wetland_quality
      quality)
IF (in wetland_quality wetqual = <wetqual>)
  THEN fuzzify 1 quality <wetqual>;
:rule r5 rule block 0
: (goal Fuzzify hydro in Hydrology runoff)
IF (in hydrology hydro = <hydro>)
  THEN fuzzify 1 runoff <hydro>;
:rule r7 rule block 0
: (goal Fuzzify beau in Aesthetics beauty)
IF (in aesthetics beau = <beau>)
  THEN fuzzify 1 beauty <beau>;
```

An alternative set of rules that can be used to calculate directly a measure of "goodness" are these:

```
: Policy 0 -- Vegetation
IF plant density is High
  AND many different species are present
  AND species distribution is patchy
    THEN ec_flora is Good.
IF native species are present
  AND those population numbers are High
  AND protected species are present
    THEN ec_flora is Good.
IF plant growth forms are Many
  AND animal habitat types are Numerous
    THEN ec_flora is Good.
IF invasive species are present
  OR noxious weeds are present
  OR most species are annuals
    THEN ec_flora is Not_Good.
IF most plant biomass is agricultural crops
  OR cattle graze on the plants
    THEN ec_flora is Not_Good
```

and,

```
: Policy 2 -- Wetlands
IF wetland area is Large
  AND wetland type is Many
    THEN ec_wetland is Good.
IF wetland plants are varied
  AND wetland hydrology is at the surface
    THEN ec_wetland is Good.
IF wetland function is flood storage
  OR wetland function is water quality
       improvement
  OR wetland function is unique animal
       habitat
    THEN ec_wetland is Good.
IF wetlands are prior-converted farmland
  OR wetlands have been drained
  OR wetlands are used for cattle grazing
    THEN ec_wetland is Not_Good.
IF wetlands must be actively managed
  OR (wetlands are isolated
  AND wetlands are abundant in the
       vicinity)
    THEN ec_wetland is Not_Good.
```

The data are entered from a simple ASCII file:

```
"vegetation",529,0,0,0,0,0
"te_plants",0.0,0,0,0,0,0
"wildlife",2.0,0,0,0
"te_anim",5.0,0,0,0,0,0
"wetland_size",138.0,0,0,0,0,0
"wetland_quality",60.0,0,0,0
"hydrology",12.5,0,0,0,0,0
"water_quality",8.0,0,0
"aesthetics",2.4,0,0,0
"traffic",55.0,0,0,0
"sig_likelihood",53,0,0,0,0,0
"sig_magnitude",12,0,0,0,0,0
"sig_area",19.0,0,0,0,0,0
```

```
"sig_duration",24,0,0,0,0,0
"sig_reverse",75,0,0,0
"sig_mitigation",75,0,0,0
"sig_timing",3,0,0
"sig_cumeffect",25,0,0,0,0,0
```

Notice that the eight components of significance are entered the same way as the rest of the data. In the model used, an initial zero (0) is required for each fuzzy term set in the linguistic variable.

The flexibility of fuzzy sets and fuzzy logic to model the underlying semantic meaning of a process as complex and sensitive to values and beliefs as an environmental impact assessment makes them a powerful set of tools that provide a quantitative, but open, answer to a subjective question.

17.2 Environmental Condition Indices

The values for the existing conditions and all five project alternatives are the input to the FuzzyEI-Assessor™ model, as partially shown above. The component weights are determined during the scoping process and represent consensus importance values. All six data sets are processed the same way, except that impact significance is not applicable to existing conditions. This procedure removes several potential contentious issues with traditional assessments and the time required is dependent on data entry and not manual computation or making value judgments that are not necessarily technically sound or legally defensible.

The ECI values are summarized in Table 17.1 on the following page and are interpreted two ways: as an example for the purpose of illustrating the approach and its application and in the real world. The limitations for interpretation of these results arise primarily from the entire assessment's not being designed from the beginning for this analytical approach. Therefore, a lot of valuable data were not available for inclusion in the example model runs. This is no different from the application of statistical analyses to data without similar forethought. Therefore, no conclusions can be drawn about alternatives specific to the Columbia Gateway site owned by the Port of Vancouver, USA because the example was too superficial to support management decisions. This points out the necessity of careful and complete planning

Table 17.1. Environmental Condition Indices (ECI) for the existing conditions and proposed alternatives at the Columbia Gateway site, Port of Vancouver, USA.

Condition	ECI
Existing	0.80
Alternative 1	0.60
Alternative 2	0.86
Alternative 3	0.78
Alternative 4	0.85
Alternative 5	0.77

before any data collection efforts are begun when the modern approach to environmental impact assessments is to be used.

Despite the lack of a lot of data and analyzing fewer than a dozen components, the ECI results illustrate both relative rank and overall "goodness" or "acceptability" of the site as it is and as it could be under various development scenarios.

The existing conditions are the third highest in the set of six, while two sets of alternatives (numbers 2 and 4 and numbers 3 and 5) are separated by only 0.01 ECI units and the "No Action" alternative evaluates far below all other conditions. The very close alternatives in each pair are essentially the same. It is not possible to make anything of such a small difference. The comparative deterioration of the site with no active mitigation or other management actions is shown to be real. In some environmental impact assessments the assumption is that doing nothing is the best thing for the environment, social values and economic conditions in the affected area. The example shows that this assumption is not always warranted. Because Alternative 5 (complete build-out of the site) was not in the actual assessment, no data were collected on the large state wildlife refuge to the north. Because that area is the recipient of mitigation efforts in this alternative there was no way to include those in the calculation of the ECI. Had these data been available the final values may well have been greatly different.

Overall, the calculated ECI values are higher and more tightly clustered than might be intuitively expected. The ECI has a range of 0.0–1.0 and all alternatives except the "No Action" are in the upper quartile.

This suggests that any of Alternatives 2–5 could be accepted with minimal overall change from the existing conditions.

17.3 Making and Supporting Policy Decisions

The modern approach to environmental impact assessment presented and explained in this book does not remove the responsibilities and obligations for people to make the final decisions. In the example, the decision-makers could justify adopting Alternative 5 despite its comparatively low index value because of the greater economic and jobs benefits, but with adjustments to bring the calculated ECI up somewhat.

Whatever they decide to do, they have both the full set of ECI values and a full audit trail of how each value was calculated by the model. Rather than being a "black box" process, the modeling is a "transparent box" process. For any (or all) alternatives and components, the intermediate consequent fuzzy sets can be printed so the effects of each rule can be seen. This review can answer the question of how a result was calculated, indicate where more (or more refined) data would be useful and, in general, support the ultimate decision.

Having objective values upon which to base a decision can be a major benefit when the project is complex, contentious and fraught with political implications for the decision-makers. The charge that decisions are arbitrary or capricious is difficult to substantiate with a transparent process that is objective and based on a solid mathematical foundation.

17.4 Caveats

No analytical method can make meaningful results out of insufficient data. This is more true for this modern approach to environmental impact assessments than even for less complex statistical data analyses.

The model structure must be defined first and used to determine the extent and composition of the entire assessment process. Because the manipulations of values, data, and components is done by computer there is no reason to limit choices arbitrarily. For example, dozens of environmental, social, economic, and political components can be identified as available for inclusion in the assessment. The number actually included can be determined after the scoping process has allowed every stakeholder and other interest group to participate in the ranking of these components. For a large, complex project (e.g., a metal mine, hydroelectric dam, or timber harvest in an area with threatened or endangered plant or animal populations), forms can be printed with all available pair-wise comparisons and the data converted to digital format by a scanner designed for that purpose. Such scanners are routinely used for grading student examinations and voting. This method reduces manual labor and the associated risk of incorrectly entering a preference to the digital database.

The modern approach permits incorporation of subjective, observational data with measured data when establishing baseline conditions and trends. Careful thought must be applied to determine the types and frequencies of observational data that can be useful to the assessment. This means that historical records can augment the data used in the analyses even when anecdotal reports are not included. These decisions must be made beforehand.

The shape of membership functions used as terms in linguistic variables, the number of terms, and the range of each can effect the outcome of the model runs. If time permits it would be advantageous to conduct sensitivity tests on existing condition data before alternatives are analyzed. These tests are similar in purpose to regression tests run on database systems. Vary input data—one component at a time—by a fixed amount (usually a percentage of the original value) and re-run the model. If the results vary from the first model run by a greater or lesser percentage than the change in input data then the model is more or less sensitive to the input data values. Similarly, with the same input values but different membership curve shapes the sensitivity to the model of minor changes can be assessed. Ideally, the analytical model will be structured so it is sensitive to changes in important values and insensitive to changes in those that are less important or more coarsely defined. Which values are considered more important, and which less important, will vary with location and type of project.

17.5 What the Future Holds

The use of fuzzy sets and fuzzy logic for environmental impact assessments is only the beginning of business and societal benefits to be gained from computational intelligence applied to environmental concerns and decisions. Other major tools briefly mentioned in this book are artificial neural networks and evolutionary (or genetic) algorithms. These two techniques can be combined with fuzzy system models to solve almost all the problems encountered that have environmental consequences.

Artificial neural networks mimic the behavior of the human brain. They are trained to recognize patterns and then turned loose on large volumes of data. They find a wide array of subtle patterns not easily seen by scanning data but they do not work transparently. Unlike fuzzy system models where everything is explicit and open, artificial neural networks come up with the right answer but with no way of determining exactly how they did it. This may not be politically acceptable in some circumstances, but heuristically they can be proven correct. In the business world, neuro-fuzzy models are used to assess and manage risks in many areas. There is no reason why these models cannot be applied to environmental risk management with the same societal and business benefits.

Evolutionary algorithms mimic our understanding of genetic mutations and transference from one generation of humans to future generations. Such computer models have been proven very efficient and correct in "evolving" optimal strategies and solutions. An ideal application that fuses evolutionary algorithms with fuzzy system models is the analysis of large data sets, particularly those collected over many years and large areas. Properly constructed and tested, the evolutionary algorithm will identify the rules applicable to a fuzzy system model. As examples, consider the streamflow data available from the Water Resources Division of the US Geological Survey and the water quality databases maintained by the US Environmental Protection Agency. These data have been collected at thousands of monitoring/sampling stations, sometimes for most of a century (for surface water flow data). If a project (for example, an environmental impact assessment) would benefit from inclusion of these data to justify the structure of rules, they would be almost impossible to analyze manually. But, an evolutionary algorithm will use starting values and evolve a set of optimal rules relating various parameters by mutating the digital "genes," passing the "most fit" to succeeding generations, and acting on the data the same way that Darwinian evolution acts on populations of organisms over many generations.

These are exciting times for those in the applied environmental sciences. All environmental laws and regulations are based on subjective values, not hard science. The methods and approaches that societies have applied over the past decades have repeatedly proven themselves either broken or barely inadequate. The tools to fix the situations have been available, and now they have been applied to environmental impact assessments. This is only the beginning.

References

[1] Anile, A. M. and Gallo, G. (2001). Fuzzy numbers for environmental impact assessment. Department of Mathematics and Information, University of Catania, Italy.

[2] Bahus, K. (Ed.) (2002). Port of vancouver, columbia gateway subarea plan draft environmental impact statement. Technical Report SEP 2002-00040, Ciy of Vancouver, Vancouver, WA 98660.

[3] Berkan, R. C. and Trubatch, S. L. (1997). Fuzzy Systems Design Principles: Building Fuzzy IF-THEN Rule Bases. IEEE Press, New York, NY.

[4] Canter, L. W. (1977). Environmental Impact Assessment. McGraw-Hill Book Company, New York, NY.

[5] Canter, L. W. and Canty, G. A. (1993). Impact significance determination—basic considerations and a sequenced approach. *Environmental Impact Assessment Review*, 13:275–297.

[6] Cornelissen, A., van den Berg, J., Koops, W. J., and Kaymak, U. (2002). Eliciting expert knowledge for fuzzy evaluation of agricultural production systems. Technical Report ERS-2002-108-LIS, Erasmus Research Institute of Managment, Rotterdam, The Netherlands.

[7] Council on Environmental Quality (1987). Regulations for Implementing NEPA, Section 1508.27, 40 Code of Federal Regulations.

[8] Cox, E. D. (1995). Fuzzy Logic for Business and Industry. Charles River Media, Rockland, MA.

[9] Cox, E. D. (1999). The Fuzzy Systems Handbook, Second Edition. Academic Press, San Diego, CA.

[10] Dee, N., Drobny, N., Duke, K., Whitman, I., and Fahringer, D. (1973). An evaluation system for water resources planning.

[11] Division for Sustainable Development (2003). Agenda 21. Technical report, Division for Sustainable Development, United Nations Department of Econmomic and Social Affairs, New York, NY.

[12] Duinker, P. and Beanlands, G. (1986). The significance of environmental impacts: an exploration of the concept. *Journal of Environmental Management*, 10(1):1–10.

[13] Environment Australia (2002). Environmental impact assessment training resource manual—second edition. Technical report, Economic and Trade Branch, United Nations Environmental Program, Geneva, Switzerland.

[14] Harrop, D. O. and Nixon, J. A. (1999). Environmental Assessment in Practice. Routeledge, London and New York.

[15] Haug, P. T., Burwell, R. W., Stein, A., and Bandurski, B. (1984). Determining the significance of environmental issues under the national environmental policy act. *Journal of Environmental Management*, 18:15–24.

[16] Klir, G. W. and Wierman, M. J. (1999). Uncertainty-Based Information, Second Edition. Springer-Verlag, Heidelberg, Germany.

[17] Larcombe, P. (2000). Determining significance of environmental effects: an aboriginal perspective. Technical report, Canadian Environmental Assessment Agency.

[18] Martel, G. F. and Lackey, R. T. (1977). A computerized method for abstracting and evaluating environmental impact statements. Technical report, Virginia Polytechnic Institute and State University, Blacksburg, VA.

[19] Marusich, L. J. (2001). The application of fuzzy logic analysis to assessing the significance of environmental impacts: Case studies from Mexico and Canada. Technical report, Canadian Environmental Assessment Agency.

[20] Mendel, J. M. (2001). Uncertain Rule-Based Fuzzy Logic Systems: Introduction and New Directions. Prentice-Hall PTR, Upper Saddle River, NJ.

[21] Mendel, J. M. (2003). Type-2 fuzzy sets: some questions and answers. *IEEE Neural Networks Society*, pages 10–13.

[22] Mendel, J. M. and John, R. I. B. (2002). Type-2 fuzzy sets made simple. *IEEE Transactions on Fuzzy Systems*, 10:117–127.

[23] Negoita, C. V. (1985). Expert Systems and Fuzzy Systems. Benjamin/Cummings Publishing Company, Inc., Menlo Park, CA.

[24] O'Hagan, M. (1990). A fuzzy decision maker. Unpublished manuscript, Fuzzy Logic, Inc.; La Jolla, CA.

[25] Reynolds, K. M. (2001). Fuzzy logic knowledge bases in integrated landscape assessment: examples and possibilities. Technical Report PNW-GTR-521, U.S. Forest Service, Pacific Northwest Research Station.

[26] Rossouw, N. (2003). A review of methods and generic criteria for determining impact significance. *African Journal of Environmental Assessment and Management*, 6:44–61.

[27] Russell, B. (1923). Vagueness. *Australasian Journal of Psychology and Philosophy*, 1:84–92.

[28] Saaty, T. L. (1977). A scaling method for priorities in hierarchical structures. *Journal of Mathematical Psychology*, 15:234–281.

[29] Saaty, T. L. (1991). Multicriteria Decision Making: The Analytic Hierarchy Process, Second Edition. RWS Publications, Pittsburgh, PA.

[30] Sadler, B. (1996). Environmental assessment in a changing world: Evaluating practice to improve performance, final report. Technical report, Canadian Environmental Assessment Agency, Ottawa, Canada.

[31] Silvert, W. (1979). Symmetric summation: A class of operations for fuzzy sets. *IEEE Transations on Systems, Man, Cybernetics*, 9:657–659.

[32] Silvert, W. (2000). Fuzzy indices of environmental conditions. *Ecological Modeling*, 130:111–119.

[33] Sippe, R. (1999). Criteria and standards for assessing significant impacts; in Petts, J. (Ed.), Handbook of Environmental Ipact Assessment, Volume 1, Environmental Impact Assessment: Process, Methods and Potential, pages 74–92. Blackwell Science, London.

[34] Sólnes, J. (2003). Environmental quality indexing of large industrial development alternatives using AHP. *Environmental Impact Assessment Review*, 23:283–303.

[35] Tompson, M. A. (1990). Determining impact significance in eia: a review of 24 methodologies. *Journal of Environmental Management*, 30:235–250.

[36] Tran, L., Knight, C. G., O'Neill, R. V., Smith, E. R., Ritters, K. H., and Wickham, J. (2002). Fuzzy decision analysis for integrated envi-

ronmental vulnerability assessment of the mid-Atlantic region. *Environmental Management*, 29:845–859.

[37] U.S. E.P.A. (1993). Sourcebook for the environmental assessment process. Technical Report 300-B-93-007, Office of Federal Activities, Environmental Protection Agency.

[38] Veiga, M. M. and Meech, J. A. (1994). Application of fuzzy logic to environmental risk assessment. In Proceedings of the IV Encuentro Hemisferio Sur sobre Tecnologia Mineral, Concepcion, Chile. University of British Columbia.

[39] Yager, R. R. (1977). Multiple objective decision-making using fuzzy sets. *International Journal of Man-Machine Studies*, 9:375–382.

[40] Yager, R. R. (1981). A new methodology for ordinal multiobjective decisions based on fuzzy sets. *Decision Sciences*, 12:589–600.

[41] Yager, R. R. (1988). On ordered weighted averaging aggregation operators in multicriteria decision making. *IEEE Transactions on Systems, Man, Cybernetics*, 18:183–190.

[42] Zadeh, L. A. (1965). Fuzzy sets. *Information and Control*, 8:338–353.

[43] Zadeh, L. A. (1973). The concept of a linguistic variable and its application to approximate reasoning. *Memorandum ERL-M 411, Berkeley, October 1973*. Cited in Zimmerman (2001).

[44] Zimmerman, H.-J. (1987). Fuzzy Sets, Decision Making, and Expert Systems. International Series in Management Science/Operations Research. Kluwer Academic Publishers, Boston, MA.

[45] Zimmerman, H.-J. (2001). Fuzzy Set Theory and its Applications. Kluwer Academic Publishers, Boston, MA, fourth edition.

Index